Contents

Acknowledgements

In revising this book I have drawn on many sources too numerous to mention, but primarily from my contributions to *Farmers Weekly*. However, I am also indebted to standard works such as *The John Deere F.O.S. Manual*, and *Plumbing 2* by A.L. Townsend.

Vic Moore

FARM WORKSHOP & MAINTENANCE

The book of the **Farmers Weekly** Series

THIRD EDITION

Revised by Vic Moore

GRANADA
London Toronto Sydney New York

Granada Technical Books
Granada Publishing Ltd
8 Grafton Street, London W1X 3LA

First published in Great Britain 1972 by Crosby Lockwood Staples
Reprinted 1978
Second edition 1979
Reprinted 1980 by Granada Publishing
Third edition 1984

Copyright © 1972, 1979, 1984 Farmers Weekly

ISBN 0 246 12019 3

Typeset by Cambrian Typesetters, Aldershot, Hants
Printed in Great Britain by Billing and Sons Ltd., Worcester

Introduction

Well-maintained machinery is the lynch pin of the successful farm business. Break-downs cost time and money in the idle time spent waiting for the dealer's fitter to arrive and the cost of the spare parts required. Regular maintenance keeps both to a minimum. Hidden costs lie in lost opportunities — the grain not harvested because the combine harvester was laid up during the two fine days before the deluge; the silage crop left down too long because the forage harvester was out of commission; the winter wheat which should have gone in by mid-October but did not because the drill was defunct.

At some time or other, it has happened to us all. That is why many of the most efficient farm units maintain a farm workshop, and with it a capacity for on-the-spot machinery repair which keeps down-time to the minimum.

Farmers Weekly believes that all farms in the UK would benefit by having an element of maintenance and repair facility, be it just a bench and a few handtools or a thorough-going workshop equipped with the latest engineering aids. That is why we offer this workshop book. It is laden with hints on how to cope with the 1001 jobs — some big, some small — that inevitably arise from having machinery on the farm. We have set out to help our readers save time and consequently money — two of the most important facets of successful farm management.

Denis Chamberlain
Editor

Field and Bench Notes

Tool-kit or telephone?

A man good with his hands can make repairs or adjustments when the need is pointed out, but may sooner or later ask himself whether major overhauls should be attempted in the farm workshop or limited to general upkeep and routine maintenance. The ability of the individual is critical. A little knowledge is as dangerous here as around any other part of the farm.

Failure of starter motors, generators and other proprietary items causes no major problems in this age of replacement units. Usually the remedy is simply a matter of obtaining the necessary part and fitting it.

The time to start thinking about drawing the line comes when complete engine, transmission, or fuel injection equipment overhauls are thought to be necessary. Before having a go at this sort of job call in the dealer and get his advice. It may be that the symptoms have been incorrectly diagnosed. For example, loss of engine power may be the result of stuck, worn rings, the injection pump not functioning normally or something as simple as an obstructed air supply.

Judgement and costs

Special tools and equipment required for carrying out major repairs on a small number of tractors are often too costly for the amount of work to be done. Some farmers with large well equipped workshops have found this out the hard way and have reverted to employing local dealers to do major overhauls. In any case the time spent on this sort of workshop job should never be at the expense of more important jobs outside, especially when the weather is right.

Clean up campaign

Farm dirt and the precision parts that go into an increasing number of farm machines nowa-days do not go well together. Makers of electronic components who thought they were supplying for an elementary process by their standards, have had to think twice when faced with the combination of dirt and damp under which their units were expected to function on farms. Now they seal them or encase them in plastic. A replacement part is simply plugged in.

Engine manufacturers have lived with the problem for much longer and have developed their own ways of keeping dirt at bay. But filters still rely on basic common sense and co-operation from the user if all is to go well.

A diesel engine contains precision parts which tolerate far less 'dirt' than many people realise if they are to run efficiently. It is difficult enough to think in terms of fractions of a millimetre and a micron may not seem worth bothering about. Yet at this size a particle of something as fine as cigarette ash or lamp black seems like half a brick to the finely ground mating surfaces of a diesel tractor's pump and injectors and can soon start wearing away these surfaces.

Look at the book

A browse through the instruction book is essential before putting any piece of equipment to work. This may sound elementary, but it is often ignored particularly if the machine is a direct replacement of a familiar model.

What is usually forgotten is that the machine is almost certain to have been modified since the original and operating and maintenance procedures may also have changed.

Many manufacturers spend a lot of time and money on their instruction books; others make do with a printed sheet of so-called instructions.

Good though the former editions may be there is always room for improvement, usually in simplifying explanations still further. Familiarity is a dangerous thing for instruction book scribes, the more they know about a machine the more difficult it is to look at the subject through the eyes of the man who has just bought one and knows nothing about it.

Comprehension test

There is much to be said for the technique of the executive in one company, who admitted that he knew little about the technical side of the equipment he sold but insisted that he saw all publicity and instruction material before it was published. 'If I can understand it anyone can,' he said.

To the firms which issue the other sort of instructions the message is clear. Apart from a moral duty to make sure that whoever buys the machine knows how to work it, the poor farmer can hardly be blamed if he does the wrong thing.

Happily, more and more firms are becoming conscious of the need to provide the right sort of operating instructions, with useful tips on how to get the best out of the particular piece of equipment in the varying conditions under which it will be expected to work.

Our idea is to aid and abet the dissemination of information. Perhaps the time will come when such a bridge between maker and user will not be necessary. Until then, we hope to help you to get more and better work done with fewer frustrations.

Calibration problem

Following up the instruction theme, the calibrating of fertiliser spreaders is a case where the farmer needs more help. In these days of increasingly concentrated fertilisers, the misses and double doses are embarrassingly evident, quite apart from the financial waste.

Many spreader makers issue clear setting instructions, but farmers ought to be able to calibrate for themselves, especially as a check after a year or two of work may be desirable to allow for normal wear.

Calibrating a spinning disc spreader to find its spread pattern across the direction of travel means putting down a line of boxes, driving over them, and tipping the results into a row of tubes which then show the 'hump' visually so that the necessary overlap can be calculated.

Demonstrations of such calibrations have shown it is necessary on some machines to cut down the makers' recommended bout width by up to a metre to get even coverage.

Some other way?

The only ways to calibrate for application rate per hectare are to sow a measured area and calculate, or put 50 kg in the hopper, run it out, then measure up and work out the calibration. This assumes that the maker's bout width measurements are correct.

There must be a more simple universal way apart from the one or two specialist versions developed for individual machines by more enlightened manufacturers.

The need for accurate calibrating becomes more necessary as fertiliser application rates and prices increase. In fact, the 'big three' fertiliser companies do offer a spreader clinic service to make sure that their regular customers have machines that use their fertilisers efficiently.

Colour guide to lubrication

Locating the position of grease nipples on new machines can, even with the help of an instruction manual, be like looking for a needle in a haystack. And remembering when they should be greased can test the best memory.

The time taken for maintenance has been reduced considerably by the introduction of sealed bearings, but some bearings still require the attention of the grease-gun. It is not always lack of lubrication that causes failures; too much grease will in many cases lead to just as much trouble.

A simple system, which helps in ensuring the correct greasing period, helps in locating the nipples and saves having to wander around with manual in one hand and grease-gun in the other, is to identify nipples with different coloured paint. Paint them in accordance with the maintenance schedule, one colour for those needing four-hourly attention, another colour for eight-hourly, and so on.

This will cut the time spent on maintenance and help safeguard against missing an important grease point. Nipples can be seen at a glance through the colour coding when they need attention.

Much time and grease can be wasted through blocked or damaged nipples. To overcome this in the field, place a piece of rag between the nipple and the grease-gun nozzle. It will act as a seal and give a greater pumping pressure. If this method fails a new nipple will have to be fitted.

The arch enemy

Rust is an arch enemy of the farmer. It attacks and eats into all ferrous metals and if not checked, will quickly reduce the resale value of a piece of tackle considerably. However, most metal surfaces can be treated to prevent corrosion by painting, galvanising or applying one of the many preparations on the market.

Nevertheless, few machines have nuts and bolts which have had their threads so treated, and the old problem of seized nuts is still with us. Do not resort to fitting a length of pipe on to the spanner, or reaching for the hammer and chisel, there are a number of remedies for releasing seized nuts.

'First aid' on stubborn nuts

Penetrating oils, obtainable in aerosol tins, have rust solvents and lubricants as part of their ingredients, and if given time to work on the rust will usually do the job. However, it is not always to hand when needed in the field. A drop of diesel in this instance will deal with rust much more quickly than ordinary oil because being thinner, it gets further into the threads.

A gentle tap on opposing flats with two hammers will usually release a stubborn nut. This breaks the rust crust between the threads and allows the oil to penetrate.

Oiling roller chains

Under some abrasive conditions chains will last longer without lubrication though the oiling of roller chains is usually advisable to increase their life. Periodic dipping in an oil-bath is recognised as the best method, but if this is inconvenient brush on or squirt the oil straight from the can.

Brown or black discoloration or red oxide deposits on the links are warnings of insufficient lubrication, which will also show up as different rates of wear in different areas of the same chain.

If a chain is removed from any assembly which involves timing make sure the timing marks on the sprocket are in the right place before it is replaced.

For some jobs double sprocket drives are used and teeth must be in line, otherwise jerky drives, sprocket wear and chain failure will result.

Sometimes 'skip tooth' sprockets, so called because every other tooth is missing, are used to prevent the crop building up between the chain links and the teeth. On driving elevator chains it is important that slats or flights in the chain are in mesh with the skipped tooth.

The right amount of sag in a chain is important. A chain allowed to whip because it is too slack puts tremendous shock loading through the pins and bushings. On the other hand a chain adjusted too tightly will prevent oil penetrating between the pins and bushings and result in rapid wear.

Generally, chain sag should be 3% of the slack length. For example, 50 cm of slack chain calls for 15 mm sag.

Safety – inside and out

A discussion on farm machinery is incomplete without some mention of safety. Though some may think that the safety message is rammed too hard down our throats these days, there can be no argument about the figures; 71 killed on farms during 1981, 13 of them children. Furthermore, 9 of these children were under 10 years of age. (Health & Safety Executive News Release May 1982.)

Though the regulations forbid a child under 13 to drive or ride on a tractor, and driving some modern tractor and implement combinations is beyond their ability, one still sees them trying.

Constant vigilance to prevent the grisly record from growing is needed in the implement shed or workshop as well as in the fields. Tackle parts removed for repair are often propped up in the most dangerous way. One sees people relying upon a comparatively flimsy jack and no secondary support to prevent a heavy implement from crushing them if something should go wrong. It all boils down to common sense, but too often this is impaired by familiarity and the contempt which accompanies it.

CHAPTER 1
Back to Basics

HORSEPOWER

Horsepower is the most misunderstood word on the farm.

The term originally meant the amount of work one horse could do in a given time. Mechanically, however, an engine with more horsepower will not always pull more — it may only go faster. A car with a 70 kW engine will not necessarily pull a four-furrow plough even if the engine was in a tractor chassis. The factor which decides how much an engine will pull is the amount of *torque* developed at the end of the crankshaft.

Torque is a rotating effort or the amount of force applied to the end of a crank arm multiplied by the length of the crank arm and measured in units called 'pounds feet'. It can be explained simply as leverage; the longer the crank arm the more work achieved for the same force. Although many readers will have come to terms with metrication, others will still think in imperial units so these are used first to explain the meaning of horsepower.

If a force of 25 lb is required to turn a socket brace with a crank arm length of 1 ft one revolution, the torque is calculated as force x leverage, 25 lb x 1 ft = 25 lb ft torque. If the crank arm length was doubled more torque would result from the same force — 25 lb x 2 ft = 50 lb ft

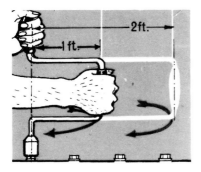

Fig. 1.1 Torque = force x leverage.

torque (Fig. 1.1). Doubling the lever doubles the torque and so on.

In an engine, the longer the crankshaft stroke the more torque is developed. A long-stroke engine has more pulling power than a short-stroke engine because the crankshaft has more leverage.

James Watt, inventor of the steam engine, found a good horse could hoist coal from a pit at an average rate of $366\frac{2}{3}$ lb at 1 ft per second or 22000 ft lb/minute using the equation $336\frac{2}{3}$ lb x 60 seconds. Watt took a few more factors into consideration and eventually arrived at a figure of 33000 ft lb/minute and called this one horsepower.

So to arrive at a horsepower figure, the speed at which the force is applied must be included — horsepower = distance load moved in feet per minute x force in lb divided by 33000.

Back to the man operating the socket brace, the distance his hand travels in one revolution is the circumference of a circle with 1 ft radius which is 6.28 ft. If he operated the brace at 50 r.p.m. the horsepower (hp) developed would be

$$\frac{50 \text{ r.p.m.} \times 6.28 \text{ ft} \times 25 \text{ lb}}{33000} = 0.238$$

Indicated hp is the force exerted on the pistons multiplied by the length of stroke and r.p.m. divided by 33000. It is a theoretical figure and gives no real indication of the actual pulling power of an engine because up to 15% can be used up in frictional losses before it gets to the flywheel. This means of expressing an engine's power is therefore not much good to the farmer as it does not tell him how much is available.

Brake hp (bhp) is the power developed at the flywheel as measured by a dynamometer on a test bed.

Figure 1.2 shows a simplified dynamometer clamped round an engine flywheel. It consists of a brake band (A) attached to a lever which is

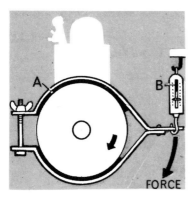

Fig. 1.2 Basic dynamometer.

in turn connected to a measuring device (B) to enable the force to be shown as lb. The same formula used to calculate the horsepower developed by the man using the socket brace is used to work out engine brake horsepower.

Brake hp is also subject to frictional losses before it eventually arrives at the point where it works for the farmer — hydraulic pumps, gear trains and various other accessories all absorb power (see Fig. 1.3). So that, as with indicated hp, not all bhp is available for work.

Pto horsepower is the actual power available at the end of the splined shaft. Though measured in a similar way to bhp, the fact that it is at the end of the shaft means that no deductions have to be made for frictional losses and it is therefore a true hp rating.

Drawbar horsepower (dhp): next to pto hp, drawbar hp is the most useful figure to have. Another type of dynamometer is inserted between the tractor and the load to be pulled.

The dynamometer shows the average pull in lb required to keep the load moving. This is multiplied by the forward speed and then divided by 33000 hp to give dhp

$$\frac{\text{load in lb} \times \text{speed}}{33000}$$

If the speed factor is in mph the formula becomes:

$$dhp = \frac{\text{load in lb} \times \text{speed}}{275}$$

The hp requirement to operate a plough demanding a pull of 2000 lb at 4 mph would be

$$\frac{2000 \text{ lb} \times 4 \text{ mph}}{275} = 29 \text{ hp}$$

Drawbar hp gives a real indication of a tractor's lugging ability. What really matters, however, is how many acres can be ploughed in a day, not how many furrows the tractor will pull. Pulling four 12 in bodies at 5 mph will result in about two acres being ploughed in an hour. If the same tractor is overloaded with six 12 in bodies working at 3 mph the work output drops to 1.8 acres an hour. The formula to use here is: acres per hour =

$$\frac{\text{working width in feet} \times \text{mph}}{10}$$

Exactly the same logic follows when metric units are used. Pulling four 300 mm bodies at 8 km/h will result in about 1.0 ha being ploughed in an hour. If the same tractor is overloaded with six 300 mm bodies working at 4 km/h the work output drops to 0.7 ha an hour. The

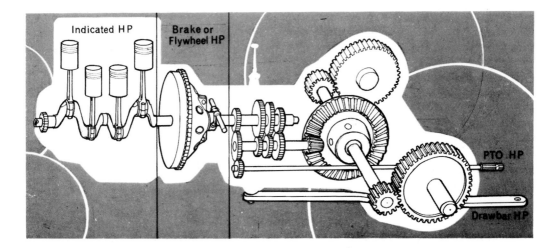

formula to use for working in metric units is: hectares per hour =

$$\frac{\text{working width in metres x km/h}}{10}$$

So when buying a new tractor or graduating to a bigger one make sure of the facts. Find out what horsepower is available for doing work and not just the power developed by the engine. When comparing the horsepowers compare like with like.

To compare engines, the power figures must be obtained under similar conditions. In order to do this, international standards exist. The main ones are:

DIN 70020 (Deutsche Industrie Normen). This is widely used in Europe and gives a power rating at the engine flywheel under the following conditions:

(a) air temperature corrected to 20°C,
(b) atmospheric pressure corrected to 760 mm mercury,
(c) fitted with *production air cleaner, exhaust, water pump and cooling fan,*
(d) *generator fitted and running but not charging,*
(e) using commercial fuel,
(f) no smoke level specified.

BS AU 141a 1971. The reference atmospheric conditions are the same as for the DIN 70020 standard i.e. 20°C and 760 mm mercury.

Ancillary equipment is *not* fitted but the air intake depression and exhaust back pressure must not be exceeded when installed.

SAE J270. This is similar to the DIN standard but the atmospheric conditions are corrected to different levels:

(a) air temperature corrected to 29.4°C (DIN 20°C)
(b) atmospheric pressure corrected to 746.2 mm mercury (DIN 760 mm)

Comparison of engine power ratings. The following illustrates the variations in power figures resulting from testing an engine to the different standards.

Standard	Power
DIN 70020	44.13 kW
BS AU 141a 1971	46.98 kW
SAE J270	45.49 kW

When comparing the power claimed, check the following:
(1) Is the power tested to a recognised standard?
(2) Is the power measured at rated speed or 'maximum' engine speed?

Rated horsepower is the brake horsepower load an engine is capable of carrying for a period of 12 hours.

Continuous horsepower is the brake horsepower load an engine is capable of carrying for continuous full-load runs of more than 24 hours.

Intermittent horsepower is the brake horsepower load an engine is capable of carrying for short periods under varying loads.

Rated, continuous and intermittent hp figures are calculated on a test bed with engines fitted with full equipment, including radiator and fan.

Rated hp is usually around 90% of maximum hp and continuous hp 90% of rated hp.

In the case of metrication, the 'horse' is dropped and we simply talk about 'power' such as brake power or indicated power and change the units from horsepower to kilowatts. To convert, the equation is:

1 horsepower is equivalent to 746 watts.

The theory of power measured in kilowatts is, obviously, exactly the same as the theory of horsepower but the units of measurement are different right down to basics. Work is done when a force is applied to a body and the body moves in the direction of the force. The amount of *work done* is measured by the product:

force x distance moved by point of application of force

Thus, if a uniform force P moves a body a distance s measured in the direction of the force, then

work done by $P = P \times s$

If the force is in newtons and the distance in metres then the units of work are *newton metres*. A unit of work equal to one newton metre is defined in the SI system as the *joule* (J). The joule is defined precisely as *the work done when the point of application of a force of one newton is displaced through a distance of one metre in the direction of the force.*

Power is the rate of doing work. In the SI system the basic unit of power is the watt (W) which is defined as a rate of working equal to one joule per second, i.e. to the work done by a force of one newton in moving through a distance of one metre in one second. Thus

$$1 \text{ watt} = 1 \text{ J/s} = 1 \text{ Nm/s}$$

You might reasonably go through the argument and then ask where the newton comes from. The newton is the SI standard of force and is defined as that force which, when applied to a body having a mass of one kilogram, gives it an acceleration of 1 metre per second per second (i.e. 1 m/s^2). So just as 'horsepower' came down to how fast a pound weight could be moved about so 'kilowatts' of power come down to how fast a kilogram can be moved. It is worth remembering that a 1 kg mass exerts a downward force of approximately 10 newtons.

GEARS

Gears play an essential part in farm machinery, varying from the high quality types used in transmission systems to the cruder cast-iron ones used in corn drills and potato diggers.

The function of a gear is to transmit power smoothly from one shaft to another, and depending on the type and number of teeth, changes in speed, torque, direction of rotation and direction of drive can be made. For example, a tractor engine speed of 2200 r.p.m. can be reduced by gearing to give a rear wheel speed of 20 r.p.m. This enables tractors to develop high torque and good drawbar pull.

Gear efficiency is controlled by the design and accuracy of machining; a carefully manufactured gear-box should transmit power with a loss of less than 2% per speed change.

Gears used to transmit heavy loads are hardened by special heat treatments and the materials used are constantly changing as new alloys are developed. For light work, case-hardened steel or just plain cast-iron gears give sufficient strength.

When a number of gears are meshed together they are known as a gear train, and a number of gear trains working in conjunction with each other make up a power train.

Calculating the speed of a driven gear is made easy if the following formula is used:

$$\frac{\text{r.p.m. of driving gear} \times \text{No of teeth in driving gear}}{\text{No of teeth in driven gear}}$$

There are many types of toothed gears and the difference between them lies in the arrangement of the teeth.

Fig. 1.4 Spur gears.

The spur, or straight tooth, gear (Fig. 1.4) is probably the most widely used and is the simplest form. The teeth are parallel to the centre line of the gear and can only be used where drive shafts are also parallel. When used in gear-boxes the teeth are chamfered to enable easier and smoother gear changing. Examples of spur gears can be found on most types of farm equipment.

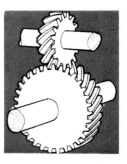

Fig. 1.5 Spiral gears.

Spiral gears (Fig. 1.5) are used to couple up shafts when the centre lines are not parallel. Contact between the teeth is confined to a single point, therefore, the load capacity is much less than that of spur or helical gears, where there is always a full line of contact between each pair of teeth as they mesh. For this reason, spiral gears are used for transmitting very light loads only.

The drive for tractormeters, speedometers and corndrill acremeters is often transmitted by this type of gear arrangement.

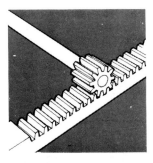

Fig. 1.6 Rack and pinion.

Rack and pinion *(Fig. 1.6)*

The rack and pinion is a variation of spur gearing, where one gear, the pinion, is a normal straight-toothed round gear and the rack is a straight-toothed flat gear. By mounting the pinion in fixed bearings and rotating it clockwise or anti-clockwise the rack can be made to move in either direction. This device is used in situations ranging from traversing a sawbench platform to adjusting the amount of fuel delivered by an in-line injector pump.

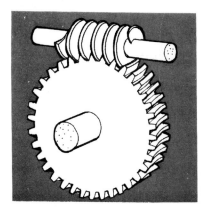

Fig. 1.7 Worm and wheel.

Worm and wheel *(Fig. 1.7)*

Worm and wheel gearing can be used for connecting shafts which are at right angles to one another and many early tractors used this method in the rear axle instead of bevel gears. The drive cannot be reversed; that is, the wheel cannot drive the worm. This made it impossible for tractors using this type of gearing to be tow-started.

Worm gears are especially useful when a big ratio of speed reduction is needed, for they can be designed to give a ratio of 70 to 1, whereas other types of gearing cannot easily give more than about 6 to 1 in any one pair.

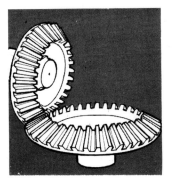

Fig. 1.8 Bevel gears.

Bevel gears *(Fig. 1.8)*

Bevel gears are used to change the course of a drive through a right angle or when it is necessary to connect two shafts that are not parallel to each other, in which case they are called angle bevel gears.

Where smooth running is required, as in the differential of a tractor, the straight cut bevel gear tends to be unsuitable and a spiral bevel gear is used. The teeth each form part of a spiral and are matched to the teeth of the mating gear. Known commonly as a crown wheel and pinion, they are matched sets not interchangeable.

The adjustment of bevel gears in relation to each other is vital. Spur and helical gears have only one adjustment, backlash; but bevel gears have two, backlash and mounting distance. If incorrectly set noisy operations and excessive wear result.

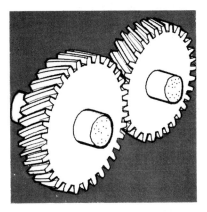

Fig. 1.9 Helical gears.

Helical gears *(Fig. 1.9)*

Helical gears are used when smooth, quieter operation is needed, and they are fitted to many modern high-speed gearboxes. Although they have great strength and operate quietly at high speed, they have a disadvantage. When under load they tend to push each other sideways out of mesh, and in some cases special thrust bearings have to be provided to prevent this.

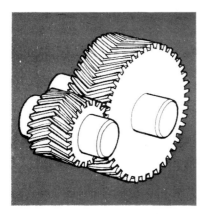

Fig. 1.10 Double helical gears.

Double helical gears overcome the problem of sideways thrust, (Fig. 1.10). The main application for this type of gear is for transmitting heavy loads.

Epicyclic gearing

Epicyclic gearing, or planetary gearing, provides a compact method of transmitting power, which may be arranged to give increases or reductions in speed.

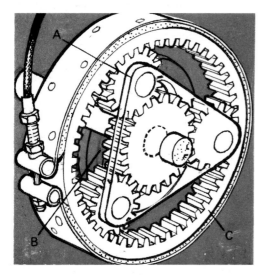

Fig. 1.11 Epicyclic gears.

The epicyclic principle is commonly used for rear axle final drive speed reduction and in transmission systems to give speed and torque amplification. Ford 'Dual Power' employs epicyclic gearing to obtain speed changes on the move without having to de-clutch.

Figure 1.11 shows the working principle of a two-speed epicyclic unit.

If the ring gear (C) is allowed to revolve the whole assembly turns as one unit, the speed of the output sun gear (B), being the same as the input planet carrier (A). When the ring gear is held stationary by a brake the planet carrier is made to rotate within the ring gear. This, in turn, speeds up the sun gear and increases the output speed.

HYDRAULICS

The power output of a tractor can be delivered at three points: drawbar, pto shaft and hydraulic system. To date tractor capabilities have been measured in terms of dhp and pto hp performance only. But with the development of sophisticated hydraulic linkage systems and hydrostatic transmissions the need for a greater understanding of how they work and how to get the best performance increases every day.

Hydraulics is the science of fluid forces and through modern usage it has come to mean the use of fluid (usually oil) to transfer or change a source of power into usable power.

In the seventeenth century a scientist called Pascal discovered the basic law upon which modern hydraulics work. The law is: if a pressure is applied at any point in a static fluid it will be the same in all directions and acts with equal force on equal areas (Fig. 1.12).

Fluids are almost incompressible and because of this forces may be transmitted, increased and controlled by means of a fluid under pressure.

All hydraulic systems, whether a simple single-acting ram for operating a combine pick-up reel or a 100 kW tractor with draught control and hydrostatic transmission, use fundamental design features (Fig. 1.13).

If a force of 1 kgf is exerted on piston A and moves it 100 mm down its cylinder, 100 x 100 = 10000 cubic mm of oil will be displaced if the area of cross section of the piston is 100 sq mm. The displaced oil will be forced against the 1000 sq mm face of ram B and lift the load of 10 kg by 10 mm. This is calculated by dividing the oil displaced by the area of the piston B.

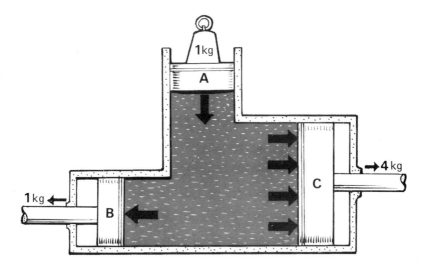

Fig. 1.12 Piston area: A = 1 cm², B = 1 cm² and C = 4 cm². Output force from B is equal to input force. The area of piston C is greater, therefore the output force is greater. Output is proportional to input.

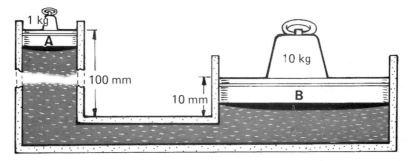

Fig. 1.13 Area of piston A = 100 mm², area of piston B = 1000 mm².

The pressure under A and B must be the same, so if B has 10 times the area it will lift 10 times as much but will do so through only 1/10 the distance.

To convert the simple systems shown above into a workable hydraulic system it is necessary to incorporate a pump, two non-return valves — one on the inlet side of the pump and one on the output side — and an oil reservoir (Fig. 1.14).

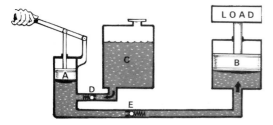

Fig. 1.14 A Handpump, B Lift cylinder, C Reservoir, D and E Non-return valves.

Any force developed by the hand pump (A) will exert a greater force on the piston (B) according to the multiplication forces (area of piston x distance piston moved).

To move the piston through a complete stroke it is necessary to have a reservoir to supply the extra oil required. Non-return valves permit the oil to be pumped from the reservoir, which is at atmospheric pressure, to the cylinder at increased pressure.

The oil pumped into the cylinder (B) cannot return to the reservoir so it is necessary to connect a directional control valve and a return pipe (see Fig. 1.15).

As most systems require a continuous oil flow the left hand pump is replaced with a motor-driven pump and a pressure relief valve (prv) is fitted (see Figs. 1.15 and 1.16).

The purpose of the pressure relief valve is to prevent damage to the pump or any other part of the system in the event of the piston

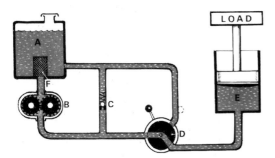

Fig. 1.15 Schematic layout of a simple hydraulic system.
A Reservoir, B Pump, C Pressure relief valve, D Directional control valve, E Lift cylinder, F Filter.

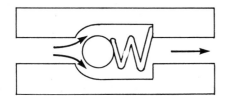

Fig. 1.16 Non-adjustable pressure relief valve.

reaching the end of its travel before the oil is diverted back to the reservoir by the control valve. Relief valves are usually set to a pre-determined pressure. In a tractor hydraulic system the internal control valve is automatically returned to its neutral position when the lower links reach the full lift position.

Ease of control is one of the big advantages of hydraulic power. The flow and pressure can be altered to give stepless speed changes and lifting capacities. It may be routed through flexible hoses or piped, thus doing away with the complicated systems required for mechanical drives.

Hydraulic power is efficient and cheap to operate, small forces are able to handle much larger loads, losses through friction are small, automatic lubrication reduces wear on moving parts, the system is very flexible in respect of the positioning of components, and speed can be easily controlled.

HYDRAULIC PUMPS

The power for a hydraulic system is supplied by the pump, and the external gear type is most commonly used for farm equipment.

It operates on the principle that as the gears revolve, oil trapped between the gear teeth (B)

and the housing is carried from the suction side (A) to the outlet side (C) of the pump, one gear being driven and the other following (Fig. 1.17).

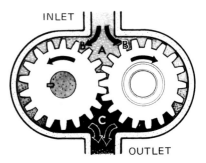

Fig. 1.17 External gear type pump.

Other types of pump include the rotor pump (Fig. 1.18) and the vane pump (Fig. 1.19).

Clearances in hydraulic pumps are critical and the smallest particle of dirt can result in operating problems. Excessive wear, for example, causes internal slippage of the oil within the pump. This is internal leakage from the high pressure side of the pump and will reduce output and increase oil temperature.

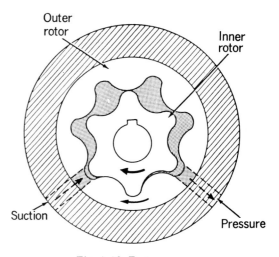

Fig. 1.18 Rotor pump.

The rotor pump operates due to the fact that the inner rotor has one less lobe than the outer rotor. Oil is drawn in as the lobes on the inner rotor slide up and out of mesh with lobes on the outer rotor. It is squeezed out again as they slide back into mesh.

Vane pumps can deliver large quantities of oil at relatively high pressures. Wear does not greatly affect efficiency because the vanes can

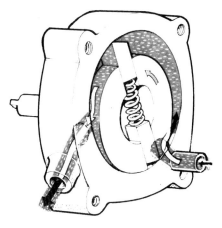

Fig. 1.19 Principle of the vane pump.

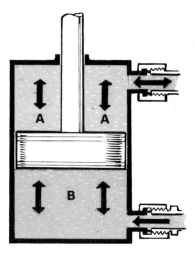

Fig. 1.21 A double acting cylinder showing how piston area A is less than area B.

move in their slots to maintain contact with the housing. But dirt can cause the vanes to seize and reduce efficiency.

HYDRAULIC CYLINDERS

Hydraulic cylinders, or rams as they are more commonly called, are most popular for converting hydraulic power into work.

They may be either single acting (S/A) (Fig. 1.20) or double acting (D/A), (Fig. 1.21) depending on the application.

A S/A cylinder relies on the weight of the load to return the piston to its home position, whereas the D/A type is under pressure in both directions.

It is important that D/A cylinders are fitted the right way round. The piston area at one end is greater than the other because of the space

taken up by the piston rod (see Fig. 1.21).

This results in a slower more powerful upward stroke and a faster less powerful downward stroke providing the pressure is the same at both ends.

The flow of oil in a hydraulic circuit is controlled by valves, of which the spool valve is the most common (Fig. 1.22). Spool valves are often mounted in banks so that several valves can share a common oil supply and return. They are very precisely made, therefore small particles of dirt entering a hydraulic system could easily damage a spool valve.

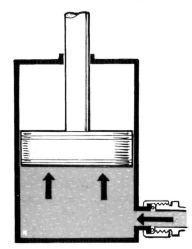

Fig. 1.20 A single acting cylinder.

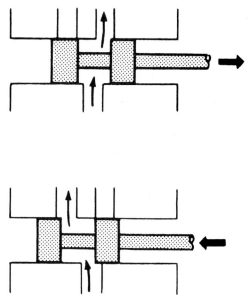

Fig. 1.22 Spool valve directing oil to one of two ports.

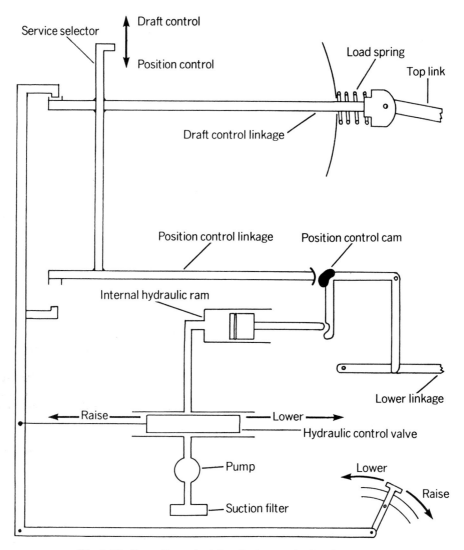

Fig. 1.23 Operating principle of a tractor hydraulic system.

TRACTOR HYDRAULIC SYSTEMS

Figure 1.23 represents a simplified hydraulic system operating in draft control and utilising the compressive force of the top link to control the depth of the implement. The system depends upon the fact that if the draft of an implement increases, then there will be a corresponding tendency for it to tip forwards (or, in the case of long implements, the tendency to hang backwards will decreased). As shown in Fig. 1.23, the forward movement of the top link acts against a strong compression spring (the load spring). With the control lever set at any particular position in its quadrant, any increase in draft will deflect the load spring a corresponding

amount and the internal linkage will therefore move the control valve to a raise position. As soon as the lower links have lifted, there will be a reduction in draft and thus the control valve will be moved back to its neutral, or lower position.

Most soil engaging implements have a tendency to pull themselves into the ground and furthermore they are usually fairly heavy. Therefore when the hydraulic system attempts to raise the lower links (due to an increase in draft being sensed through the top link) it is often easier for the front of the tractor to leave the ground than for the implement to lift. By balancing the amount of lift against additional weight on the front of the tractor, this effect can be harnessed to produce weight transfer, and increase the weight — and hence traction — at the rear wheels.

Most hydraulic systems still use the movement of the top link as a sensor for draft control. However a bottom link sensing system is, in fact, more sensitive than one employing the top link as the sensor, especially when long heavy implements are being used. This system employs the tension in the lower links instead of the compression of the top link, in order to operate the hydraulic control valve. It is usually found on tractors of higher horsepower which are likely to be operating heavier implements.

Position control

Position control is mainly used for implements such as sprayers which must operate at a constant height above the ground; once the height of the lower links has been set, they will be maintained at this level. Reference once more to Fig. 1.23 illustrates how this is achieved. When operating in position control, the control valve is acted upon by the movement of a linkage which in turn is controlled by a cam mounted on the rock shaft. If the lower links drop slightly the cam on the rock shaft will move the control valve to the raise position until the original height of the lower links is restored.

External services

In order to obtain oil under pressure from the external tapping it is usually necessary to divert oil from the internal ram which lifts the lower links. On many tractors the lower links must

not be in their highest position because the position control cam moves the control valve to neutral. The two most common causes of an external hydraulic service refusing to deliver oil are that the oil diverter valve has been accidently moved or that due to maladjustment of the system or an oil leakage the lower links have crept up.

Systems in which the oil is dumped back into the reservoir when it is not required are known as *open centre* and those which allow the pump to build up full pressure then stop pumping are called *closed centre*. The open centre type of system is much more common in the UK.

TRACTOR ENGINE GOVERNORS

An engine governor maintains a constant speed under varying loads and protects the engine and ancillary equipment against high speeds when the load is reduced.

By controlling the supply of fuel to the engine it controls the speed, which is important when driving pto tackle requiring an exact speed.

There are three basic types of governor: centrifugal (sometimes called mechanical), pneumatic and hydraulic.

The centrifugal is the type most commonly used in agricultural diesel engines (Fig. 1.24).

Two or more weights (A) are pivoted on a carrier or spindle (B), which is driven at the same speed as the engine camshaft through the timing gears. When the engine is running, the weights,

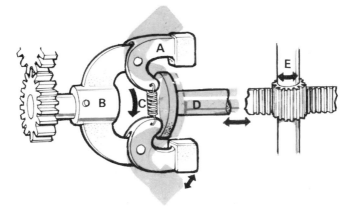

Fig. 1.24 Centrifugal governor — on many modern tractors speeds are varied from about 600 to 2500 r.p.m.

which are hinged at one end, are free to fly outwards or inwards, according to the speed at which they are driven. Their position is also controlled by a spring (C).

As the engine speed increases or decreases, the weights move in and out and this movement which is transmitted to the fuel control (E) by a push-rod (D) varies the amount of fuel delivered to the injectors.

The hydraulic governor is less widely used but it is just as efficient. It utilises the action of a spring against the working pressure of the fuel generated by a feed pump, inside the injector pump, this pressure varying with engine speed.

A shuttle valve, with spring pressure acting on one side and fuel pressure on the other, is connected to the fuel delivery valve. As the engine speed increases so does the fuel pressure and the shuttle valve is pushed against the spring to reduce fuel delivery. When the engine speed decreases the reverse action takes place and the fuel delivery is increased.

Faulty governors cause lack of power, poor engine response to increased load and 'racing' when the load is decreased.

Most governor trouble can be traced to lack of lubrication, worn or loose parts or weakened springs. If the bearing surfaces become dry and grumpy, the governor will be slow to respond to speed changes and if parts are excessively worn it may jam open or become spasmodic in operation. 'Hunting' is the term given to governor action when it produces a fast-slow-fast effect and this is usually most noticeable at slow engine speeds, when a small throttle movement makes a big difference.

The adjustments are factory set and sealed and should not be tampered with. Some drivers 'open up' the governor and say the tractor performs better. This is a fallacy and does more harm than good. Low idle setting is the only adjustment necessary. For other problems call your dealer.

LUBRICATION

The main object of an engine's lubrication system is to reduce the amount of friction and wear in the bearings and other moving parts.

No matter how smooth a bearing surface may be, a lack of oil to lubricate it will result in rapid wear, overheating and loss of power.

It does its job by maintaining a thin film of oil between rubbing surfaces, such as big and small end bearings and on the piston rings sliding up and down inside their cylinders, and by cushioning the combined sliding rolling action of gear teeth. Oil forms a seal between the piston rings and cylinder walls to reduce compression losses and also helps to keep the engine cool.

Engines with forced circulation systems are fitted with a pump which sucks oil from the sump and delivers it under pressure to all bearings and by spray or drip feed to the other assemblies.

The pressure at which the oil circulates depends largely on the type and design of the engine. This is usually around 3.5 to 4.0 bar and high-low pressure limits are usually shown on a gauge.

The pipes and borings which distribute the oil vary, but the basic principles are always the same. The drawing shows a forced circulation oil system applicable to most engines (Fig. 1.25).

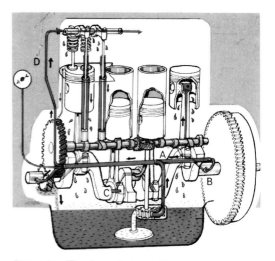

Fig. 1.25 The forced circulation system, shown in the drawing, is necessary for high speed engines to give more positive oil distribution.

The oil picked up by the pump is first fed into the main gallery (A) then routed by separate pipes or borings to the main crankshaft bearings (B). Next it is pumped through oilways bored through the centre of the crankshaft to each big end bearing (C) and oil surplus to requirement is forced out as a fine spray to lubricate the cylinder walls. A small hole drilled in the connecting rods from the big end bearings gives the small ends their supply.

The rocker shaft and valve gear are fed by a

pipe (D) leading from the main gallery (A). This pipe usually has a restrictor valve to control the amount of oil pumped to the rocker chamber and prevent flooding the top of the engine. As this oil drains back into the sump it is directed on to the camshaft timing gears and push-rods.

It is not critical to keep an exact oil level in a forced system, but it should be checked after a day's work.

If the oil gauge registers a reading above or below normal operating pressure when the engine is warm, stop immediately. On a cold morning the gauge will read slightly higher than normal because the oil is thicker.

By keeping the engine clean a leak is easily seen and the fault remedied. A leak at the rate of a drop every half-minute, which may not be obvious, will cost about a gallon of oil a month.

An engine must have good quality clean oil if it is to perform satisfactorily. If metal particles, carbon deposits and dust are allowed to accumulate they mix with the oil and form an abrasive compound, which increases the wear in bearings and piston rings.

Once rings become worn their effectiveness is reduced, causing poor compression and inefficient fuel burning. This further increases contamination and is the beginning of a vicious circle.

Generally speaking, the life of an oil is affected more by dirt contamination than by any other factor, so it is necessary to have an efficient filter — full-flow or by-pass — between the pump and the lubrication system.

The full-flow version filters all the oil delivered by the pump and discharges it direct into the system, (Fig. 1.26), whereas the by-pass takes only a small amount at a time — 8—10% — and returns it to the sump.

By-pass filters, on the whole, remove a larger proportion of dirt from the oil than full-flow filters, and many engine manufacturers prefer them (Fig. 1.27).

Because the full-flow has to handle more oil, it is usually larger and coarser.

Material used to make filters includes cotton, felt and paper and it is false economy to try to clean them, just as it is to extend the recommended oil change period.

When changing a filter element avoid dislodging the deposits inside it into the oil system.

The pump, the power unit of the lubrication system, has to supply a constant flow of oil at correct pressure into a network of pipes and oilways.

It is generally fitted low enough in the sump to be completely submerged in the oil and is driven by a vertical shaft and spiral gears from the camshaft.

Modern pumps are designed to deliver more oil than is required. Flow and pressure are

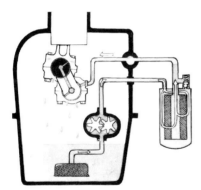

Fig. 1.26 All the oil delivered by the pump is filtered.

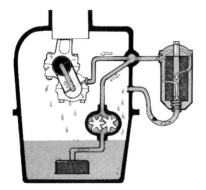

Fig. 1.27 Only a small amount of oil is filtered at a time.

Fig. 1.28 The gear pump.

Fig. 1.29 A gear pump in situ showing the prv spring gears and hinged filter screen.

controlled by a pressure relief valve (prv), usually incorporated in the pump body. The valve is pre-set to open as soon as the designed pressure is exceeded, and adjusts automatically, so that the pressure will not be affected by normal wear.

To maintain constant pressure a positive delivery pump is necessary and most commonly used is the gear and rotary-vane type.

The gear pump (Figs. 1.28 and 1.29) consists of a driver spur gear (A) meshing with an idler gear (B) within a close-fitting housing. The oil sucked in is carried round by the teeth and pressurised when they mesh, thus being forced out through the delivery port (C). The amount of oil delivered is dictated by the speed at which the pump is driven.

The vane pump (Fig. 1.30), is not as widely used. It consists of a rotor (A) which turns inside an eccentric housing (B). Two or more

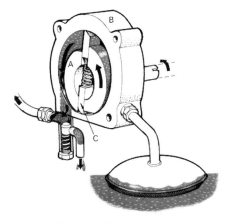

Fig. 1.30 The vane pump.

spring-loaded vanes (C), depending on the capacity required, are slotted into the rotor and, as it revolves, the ends of the vanes wipe against the housing walls. Oil is sucked in during one half revolution and pumped out during the other half.

Dust and grit get into the oil by way of the air intake and crankcase breather pipe. These small particles tend to be held in suspension in the oil and, if not removed, will cause wear to the pump mechanism and block up the oilways. It is, therefore, necessary to insert a filter.

The oil pump inlet usually has its own strainer, which is purely a primary filter of fine mesh gauze. The pump filter was at one time a gauze tray, but on modern engines it is more likely to be saucer shaped and mounted to a floating intake pipe, which rises and falls with the oil level. This draws oil from just below the surface and not from the bottom of the sump, where most of the sediment collects.

To say that oil makes a bearing surface slippery and reduces friction is only half the answer. A crankshaft, for example, floats within its bearing just as a boat floats on water. Remove the oil and direct metal to metal contact will soon wear it out. Oil must also be capable of coping with other factors and chemical changes (Fig. 1.31).

One problem facing lubrication engineers is 'cold-start corrosion'.

Fuel burned within an engine produces a considerable amount of water, some of which is exhausted as steam or in drops. The remainder mixes with the combustion gases and forms acids which eat into the cylinder walls. Engine wear is greater just after starting. When correct working temperature is reached the acids cease to be dangerous, being vaporised and exhausted.

Oxygen combined with oil, especially in mist form in the crankcase, causes oxidation, discolouring and thickens the oil. The action forms carbon and soot which gets into and thickens oil and forms sludge. This picks up fine particles of metal and dust and if oil is not changed periodically oilways may become blocked and wear speeded up. Dirty oil can cause sticking piston rings and damage to valves, leading to excessive fuel consumption. Additives help.

Additives

Additives — tailor-made chemical compounds added to an oil to improve an existing property or give it an additional one — include:

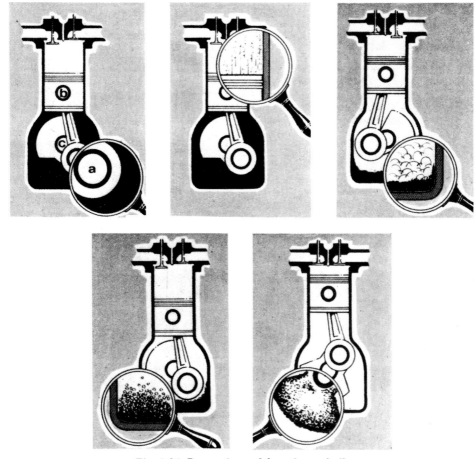

Fig. 1.31 Properties and functions of oil.

Anti-oxidant. Combats oxidation by preventing oil combining with oxygen to form harmful materials.

Detergent dispersant. Keeps engines free from lacquer deposits which form at high temperatures and sludges which form at low temperatures. Impurities and fine particles are kept separate by suspending them in oil.

Anti-wear. Improves lubrication by increasing oil's resistance to being pushed out of a bearing when under pressure.

Anti-rust additives. Beneficial where engines are not used regularly and are therefore prone to rusting. Provide protective coating on engine interior.

Anti-foam. Prevents oil foaming in crankcase.

Viscosity

Viscosity Index (VI) Improver. Viscosity index measures how much an oil thins when heated.

High VI oils thin more than low VI oils. A VI improver is inactive when the oil is at low temperatures but slows down the thinning process at high temperatures (see graph Fig. 1.32).

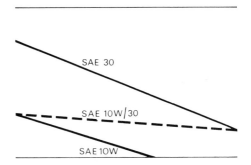

Fig. 1.32 The continuous lines on the graph show the viscosity of SAE 10 W oil at 60°C and a heavier SAE 30 oil at 99°C. The broken line shows how the thinning process is slowed down when a VI improver is added thus making a multigrade oil SAE 10 W/30.

Viscosity is measured by a viscometer, which consists basically of a metal cylinder with a small outlet in the base and mounted in an oil bath. Oil is heated to a specified temperature and the time taken for a measured amount to run through the hole is recorded.

Oils are mainly sold by a viscosity range or specification. The most commonly used specifications are those laid down by the American Society of Automobile Engineers (hence the A.S.A.E. numbers). The Americans still do not use metric units widely and the A.S.A.E. tests are not carried out in metric units. If and when the Americans do metricate, the specifications will remain the same, only the laboratory tests will change. We will still use, for example, a 20 W/50 oil for a typical engine oil or an E.P.90 as a typical gear oil.

Lubricant storage

Get round the problem of keeping tractor lubrication equipment, oil and grease in one place and available when needed by making a simple rack with a few lengths of angle iron and pieces of plywood.

Brackets made of sheet steel and bolted to the plywood back hold the grease-guns and oil-cans. Store lubricants in their containers on the floor by the rack (Fig. 1.33).

A further refinement shown in Fig. 1.33 is a built-up drip tray below the welded mesh to prevent oil staining the floor.

Fig. 1.33 Storage of oils.

PLAIN BEARINGS

Most machines have plain bearings, which require less space than ball or roller bearings and are nearly always cheaper. Plain bearings can be divided into split bushes or sleeves.

A tractor-engine connecting rod usually has a split bearing for its big end and bush or sleeve for its small end (Fig. 1.34).

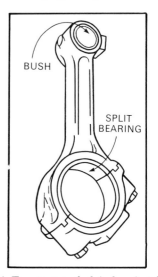

Fig. 1.34 Two types of plain bearing. The two halves of the split bearing are often called shells.

The two halves of the split bearing are often called shells. These bearings are manufactured exactly to the size required and, although care is needed during replacement, the fitting procedure is fairly straightforward and few special tools are required.

Bushes can be more difficult to replace.

The material for a bush depends upon the speed of the shaft, the load involved, what the shaft is made of, its expected use and cost. Bushes on farm machines are likely to be made of copper, brass, nylon or bronze, which is the most common.

Where the bush must operate with little or no additional lubrication, a sintered type is often used.

This so-called 'oil-less' bearing is made by compressing a mixture of powdered copper, tin and graphite. The compressed mixture is then heated to about 800°C and quenched in oil. This produces a sponge-like structure which soaks up oil. When such a bearing starts to heat up due to lack of lubrication, oil is squeezed out of its pores.

Self-lubricating sintered bearings are used where it is undesirable or difficult to apply grease or oil in any quantity, such as in dynamos, starter end plates and clutches.

Bushes are often located in small cast-iron or aluminium housings which are likely to break when struck or subjected to excessive pressure. When pressing or drifting-out a bush you should ensure that the housing is adequately supported.

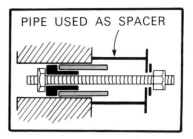

Fig. 1.37 Pulling a bush using a short hollow removing tool and puller. The tool and puller can also be used to insert the new bush.

Figure 1.36 shows how the bush-removing tool should be used. Large or tight bushes are best removed using a short hollow removing tool and a puller (Fig. 1.37). This hollow tool and puller arrangement can also be used to insert the new bush.

Sometimes bushes can only be driven out the way they went in and an adaptor sawn down each side is the best tool to use. This can be inserted through the bush, turned through 90 degrees and used to drive or pull the bush out.

Bushes located in blind or closed holes cannot be pressed out. Use the following methods to remove them.

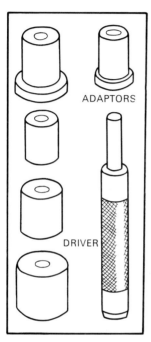

Fig. 1.35 Bush removing tool and adaptors. The tool should be a fairly tight fit inside the bush.

A special bush-removing tool should be a fairly tight fit inside the bush and have a shoulder with which to grip (Fig. 1.35). When this tool is made on the farm, it should have the edges of the shoulder square or slightly undercut, so the tool does not jam inside the bush and cause the housing to split.

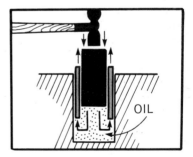

Fig. 1.38 Hydraulic removal of a bush. With luck the oil pressure will be sufficient to force the bush out.

Hydraulic method: Sometimes the bush and housing may be filled with thick oil or grease and a close-fitting shaft driven into the bush (Fig. 1.38). With luck, the pressure of oil under the edges of the bush will be sufficient to force it out. Often the shaft which normally runs in the bush can be used to remove it. This will not work when the bush is worn or has lubrication grooves the length of the bush.

Threading the bush: When the bush has a fairly thick wall, it may be possible to tap it with a suitable thread so that a screw and puller can be used to pull it out (Fig. 1.39).

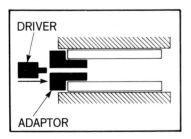

Fig. 1.36 Driving out a bush using a special bush removing tool. Bushes on farm machines tend to be made of copper, brass, nylon or bronze.

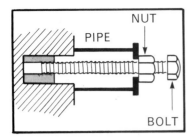

Fig. 1.39 Threaded bush being pulled out. When the bush has a thick wall, it may be possible to tap it with a suitable thread.

Bush-removal tool: Special bush-removing tools, like large self-tapping screws, have a buttress thread which gives them extra strength in the direction required. They are screwed into the bush and the bush is pulled out.

Splitting the bush: Thick-walled bushes can be split by drilling. Or a thin steel plate can be ground, hardened and tempered, and used to cut two slots in the bush. A standard cold chisel is not suitable. A small section should be cut out first as this is easier to remove.

Pullers: Occasionally the legs of a puller can be inserted behind the bush before pulling it out.

Slide hammer: A special tool can be inserted behind or into the bush, and a slide hammer used to knock out the bush (Fig. 1.40). This can easily be made on the farm.

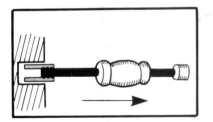

Fig. 1.40 Typical slide hammer in action. The tool can easily be made on the farm.

Sintered bushes should be soaked in oil for at least 10 hours before fitting. Always use a bearing-fitting tool. The pilot portion of the tool should be only two-thousandths to five-thousandths of an inch smaller than the installed diameter of the bearing. This will prevent bearing collapse during installation. The bearing must be driven straight into its housing. If it is 'cocked' when inserted, it may distort.

Whenever possible, use a press to insert the bearing.

Sometimes the bearing can be pulled into place using a nut and bolt and several large washers. The bore into which the bearing is to be fitted must be clean and free from burrs.

Some bushes have oil holes which must line up with a hole in the housing. When the holes are not in line, another must be drilled or the bush must be removed and repositioned.

Forcing a bush into its housing causes slight shrinkage so that the hole in the bush is reduced by about one-thousandth of an inch for each inch of its diameter. This shrinkage may not be symmetrical and if the shaft is a close fit in the bush the hole must be trued and expanded to its correct final size with a suitable reamer.

A bush may occasionally become loose in its housing. Where the bush is otherwise a good fit on its mating shaft, place a layer of soft solder on the outside, or put it in a lathe and knurl the outside. Knurling — producing a series of raised dots on the surface of a bush and the solder — increases the effective diameter of the bush and ensures a good fit.

TRACTOR FUEL SYSTEMS

The fuel system supplies a tractor's life blood at the right time and in the correct quantity — if it is working efficiently. If it is not, output suffers and tempers get frayed.

A drop of water or a speck of dust is enough to upset the balanced supply and the trouble may be found in the supply tank, the injectors, the pump, the filter, or anywhere in between.

Fuel systems can vary in layout and design according to the make of tractor but all have an injector pump incorporating some form of governor, injectors, feed pump, filters and/or a water trap and a fuel reservoir.

Repair or adjustment of the more complicated mechanisms like the injectors or pump is not usually a job for the farm workshop, but keeping the rest of the system in trim is.

The importance of clean fuel is not always sufficiently appreciated. Dirt, particularly fine dust carried in the air, is the main enemy of the diesel engine.

Tractors often work in 'dust-bowl' conditions and trouble-free operation depends on clean fuel. The injection pump, the 'heart' of an engine, and the injector nozzles which spray the fuel into the cylinders are precision made with high-finished surfaces. Abrasion from dust and dirt in the fuel can soon ruin their efficiency.

Fig. 1.41 Rotary pump.

Fig. 1.42 In-line pump.

As little as 0.003 mm or 3 microns wear of the mating surfaces in critical parts can stop it working.

A point to remember is that the manufacturer's warranty does not apply if failure is due to dirt or water getting in.

Regular day-to-day maintenance pays. The first thing a good horseman did when he finished work for the day was to rub down and feed his horse. Why not treat the tractor with the same care?

Do not drive the tractor into its shed and leave it until next morning when time for maintenance cannot be spared. Do it then. Keeping the fuel tank full helps prevent the build up of condensation. Even so a certain amount of condensation will form so keep an eye on the water tap and drain it regularly.

Look for fuel leaks. This may mean giving the engine a rub down, for leaks often go undetected because the engine and injection equipment are covered in dirt. Finally make a note of anything that occurred during the day that might mean trouble.

Injection pumps

The injection pump meters fuel and injects it at high pressure into the combustion chambers at the correct time. Both types of pump in use today — distribution (rotary) and in-line — produce the same result (Figs 1.41 and 1.42).

The pumping action of the rotary type is produced by two plungers (A) worked by a cam ring. As the rotor (B) turns one of the inlet ports (C) opens and fuel enters the rotor separating the plungers (Fig. 1.43). As the rotor turns the inlet port closes, the distributor port (D) comes into line with one of the fuel outlets, and the plungers are forced towards each other to form the injection stroke (Fig. 1.44). Fuel trapped between the plungers is forced back through the rotor to the injector.

When the plungers reach the limit of inward travel the distributor port closes, sealing off the fuel line to injector and the cycle repeats itself, sending fuel through each outlet in turn. This pump can be fitted horizontally or vertically and is lubricated by the fuel.

The in-line type has separate elements for each engine cylinder instead of the single pump barrel and plungers of the rotary type. Each element (Fig. 1.45) has a cam-driven plunger sliding in a barrel sealed at the top by a spring-loaded delivery valve. The inlet port (A) is connected to a gallery running the length of the pump, which is kept full with fuel at a constant pressure. Opposite the inlet port is a spill port (B) through which excess fuel can drain away.

As the plunger (C) descends the ports open and fuel flows into the space above the plunger, which pressurises it as it rises. The delivery valve (D) is forced off its seat and fuel flows at high pressure to the injector. This continues until the upper edge of the helix (E) in the plunger is open to the spill port, when pressure is suddenly released by fuel escaping down the groove in the plunger and out through the spill port, the

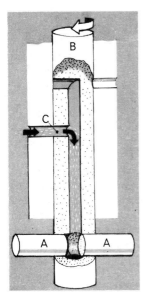

Fig. 1.43 Rotary pump — induction stroke, plungers separated.

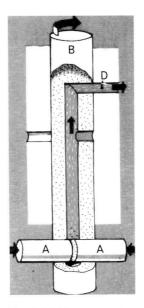

Fig. 1.44 Rotary pump — injection stroke, plungers together.

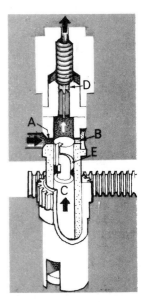

Fig. 1.45 In-line pump element — starting injection stroke.

delivery valve closes and injection ends.

The injector pump is a complicated piece of equipment and repairs and adjustments should be left to the experts, who have all the up-to-date information and special tools.

Injectors

Power output of a diesel engine is governed to a great extent by the efficiency of the injectors. Before maximum power can be obtained from the fuel injected into the cylinders, it must be atomised to allow it to mix freely with the air. This is the job of the nozzles fitted to the end of each injector.

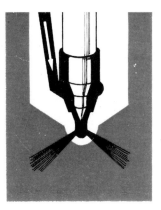

Fig. 1.47 Multi-hole nozzle.

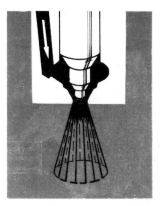

Fig. 1.48 Pintle-type nozzle

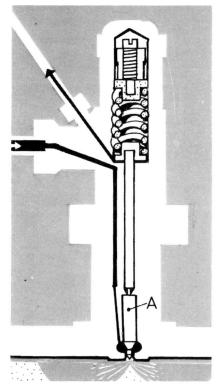

Fig. 1.46 Working principles of the injector.

Figure 1.46 shows the working principle of the injector. Fuel, under pressure from the pump, forces the spring loaded needle valve (A) from its seat and emerges as an atomised spray. As soon as the delivery pressure falls off, the spring returns the valve to its seat.

Various types are used, depending on the design of engine, the two main types being hole nozzles and pintle nozzles.

The hole nozzle (Fig. 1.47) may have two,

three or four equally spaced holes through which the fuel is forced out in fine sprays. These holes can just be seen by the naked eye, which underlines the need for clean fuel. One blocked hole can cause a power loss of up to 2 kW.

The pintle nozzle does the same job, but instead of the fuel being sprayed out of a number of holes, it is emitted as a single conical spray (Fig. 1.48).

Hole nozzles normally are used for direct injection engines and pintle for indirect. They have to withstand high pressures ranging from 100 to 200 atmospheres (100 to 200 bar) and from 150 to 1500 injections a minute, depending on the rated engine speed.

Symptoms of inefficient operation can be intermittent misfiring of one or more cylinders; smoky exhaust indicating an injector discharging unatomised fuel; pronounced knock in one or more cylinders; increased fuel consumption; engine overheating or loss of power.

Fuel lift pump

The lift pump forces fuel from the fuel tank to the injection pump via the filtering system. Usually these pumps push fuel through the filters, but occasionally they suck.

Most tractor lift pumps are mechanical, as opposed to the electric type often fitted to cars and vans. Mechanical pumps are usually bolted to the side of the engine and take their drive from the camshaft. Figure 1.49 shows the working parts. This type, with rubber diaphragm as the pumping element, is the most common mechanical pump in use today.

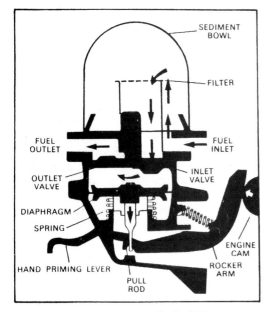

Fig. 1.49 Working parts of a fuel lift pump.

The rotating cam lifts and lowers the rocker arm which, in turn, pulls the diaphragm up and down. As the diaphragm is pulled down, the inlet valve is sucked open and fuel is drawn into the pump. When the rocker arm allows the diaphragm to lift and straighten out, the inlet valve is pushed shut and fuel is forced out through the outlet valve.

The valves are flat discs shut by light springs. Inlet and outlet valves are usually identical, but are mounted the opposite way round to one another. A return spring under the diaphragm helps it to straighten out on the discharge stroke.

Some pumps can be operated by an external hand priming lever which moves the diaphragm up and down to pump fuel when the engine is not running — as, for instance, when bleeding a diesel system.

There are several causes of trouble:

(1) Loose pump: This shortens the effective stroke and reduces output. Look for a tell-tale oil leak around its joint. Re-tighten the pump after replacing the gasket.

(2) Blocked filter: Nearly all these pumps have a filter (Fig. 1.50). To get at it, remove the inspection cover (possibly a glass bowl) which also acts as a sediment trap.

This filter should be cleaned as part of the engine's service schedule and trouble should not arise. But when contaminated fuel has been allowed into the fuel tank, there is a good chance that the filter will be blocked. Check the gasket between the pump and its inspection cover or sediment bowl. A fault here could create a fuel leak, allow air to be sucked into the pump and cause air-locks in the system.

(3) Punctured diaphragm: This will tend to reduce the efficiency of the pump, and a holed

Fig. 1.50 Lift pump filter.

diaphragm can also allow air into the system. It may permit fuel to escape through a small drain hole in the bottom of the pump which gives a warning that the diaphragm is faulty. But some pumps have no drain holes or the drain hole may be blocked.

Then the leaking fuel may find its way into the engine crankcase, dilute the lubricating oil and lead eventually to more trouble. When an engine stops using lubricating oil for no apparent reason or the level of oil actually starts to rise, suspect a leak in the lift pump diaphragm.

(4) Stuck valves: Occasionally, the pump's valves will fail to close properly. The result is reduced or no output with annoying symptoms such as the need to bleed the fuel system every day.

This is likely to happen to a combine engine

where the fuel tank is a considerable distance below the pump. When the engine is stopped there is a tendency for the fuel to drain back into the fuel tank.

A pump with a hand priming lever can be checked while on the engine. Other pumps have to be removed and the rocker arm used. Take care not to lose any gaskets or distance pieces which may be fitted between the pump and the engine. The thickness of these affects the stroke of the pump.

To check the pump, place a moistened finger over the inlet and operate the priming lever or rocker arm. A suction should be felt on the finger which should remain for a few seconds after pumping has ceased.

When repeated with a finger over the outlet, a positive pressure should be felt.

Should the priming lever on an engine-mounted pump appear to have little or no effect, it may be because the engine has stopped in such a position that the rocker arm has already pulled the diaphragm down. Turn the engine slightly to allow the diaphragm to lift.

When little or no suction or pressure is felt, the pump is faulty.

METALS

Modern machinery is made from many different materials. The farm mechanic does not have to be a metallurgist, but he often has to ask himself 'What will happen if I hit, bend or heat this component?'. It is also essential to know what material a broken part is made of, to decide whether it can be repaired by welding, or to find out what precautions are necessary.

Identifying ferrous metals

Often the appearance of a metal will give a clue to its composition. Cast-iron usually looks dull with a rough surface and may have casting marks. Dropping the components onto a concrete floor may also help: cast and malleable iron give a dull note, while good steels tend to ring. If the component is broken, examine the fracture. Cast-iron will have large crystals, either grey or white. Steel will have a fine grey crystal structure, but if it has bent many times before fracture, the metal at the break may well appear to have a large crystal structure.

One of the best tests to identify a ferrous metal is the spark test. The metal is held lightly on a grindstone, the shape, colour and length of the sparks produced varies in relation to the carbon content and heat treatment of the metal (Fig. 1.51). Another test is to try filing the metal. White cast iron will be very hard, while the steels will become harder to file as the carbon content increases.

A final clue to the type of metal is the function of the component (table 1.1).

Table 1.1 Typical materials for manufacture of farm components

Component	Mild or medium carbon steel	High carbon or alloy steel	Cast-iron	Malleable iron
Pipe (all kinds)	x			
Machinery frames	x	x		
Forgings, bolts and rivets	x	x		
Sheet metal, fenders, etc.	x			
Ploughshares, cultivator feet		x	x	
Harrow discs, hayrake tines		x		
Rough gears and sprockets, mower wheels, bearing brackets, etc.			x	
Engine blocks and cylinder heads, transmission cases			x	
Cultivator foot-piece parts, beam brackets and castings subject to shock				x
Car and tractor parts, gears, axles, shafts		x		
Springs, bearing races, etc.		x		

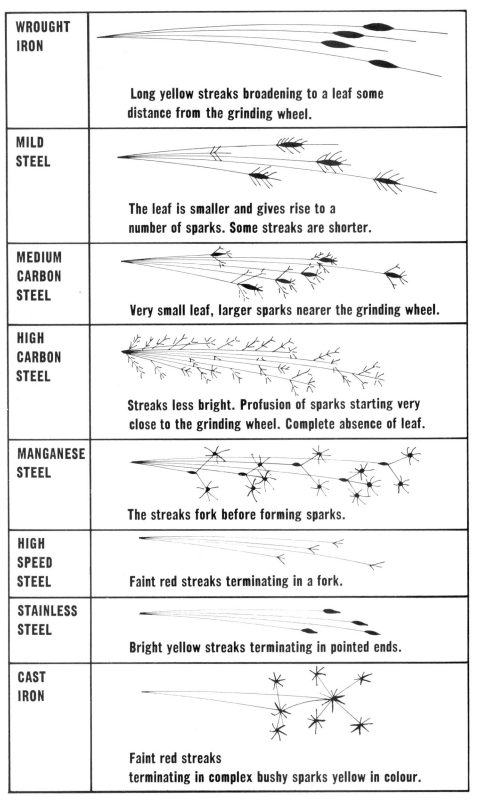

WROUGHT IRON	Long yellow streaks broadening to a leaf some distance from the grinding wheel.
MILD STEEL	The leaf is smaller and gives rise to a number of sparks. Some streaks are shorter.
MEDIUM CARBON STEEL	Very small leaf, larger sparks nearer the grinding wheel.
HIGH CARBON STEEL	Streaks less bright. Profusion of sparks starting very close to the grinding wheel. Complete absence of leaf.
MANGANESE STEEL	The streaks fork before forming sparks.
HIGH SPEED STEEL	Faint red streaks terminating in a fork.
STAINLESS STEEL	Bright yellow streaks terminating in pointed ends.
CAST IRON	Faint red streaks terminating in complex bushy sparks yellow in colour.

Fig. 1.51 The spark test.

HEAT TREATMENT

There are four main heat treatments for carbon steels: normalising, annealing, hardening and tempering.

- Normalising restores the steel to its normal condition after it has been stressed by bending or shaping during component manufacture. The component is heated to between 730°C and 900°C (red hot) for about a quarter of an hour then allowed to cool slowly.
- Annealing is similar but the component is heated for several hours then cooled as slowly as possible. This completely softens the steel.
- Hardening is carried out by heating the steel to about 750°C (cherry red), then quenching it in water or another liquid such as oil or brine. The hardness achieved depends mainly upon the quenching liquid used. Oil does not produce such a hard steel as water or brine. Fully hardened steel is extremely brittle and has a very poor shock resistance. So some of the hardness must be removed by tempering.
- Tempering consists of reheating the metal to between 220°C and 300°C depending upon the component. Then it is requenched. The temperature of the steel can be gauged fairly accurately by polishing the surface before heating and then judging the colour of the oxide film which is formed on the surface after heating. This varies from very light straw at about 220°C to dark blue at about 300°C.

Sometimes a component needs a hard surface but a tough core. One way of achieving that is to case harden. This converts the outer skin of a mild steel to a high carbon steel which can be hardened. There are several ways of introducing extra carbon into steel, but on the farm the only really practical method is to use a special compound such as Kasenit powder. The metal is heated to bright red and then dipped in the powder, which will adhere to the surface. The item is then reheated to bright red and quenched in cold water. If the case hardening needs to be deeper, then the article can be reheated and dipped into the compound two or three times before quenching.

Making use of scrap components

Table 1.1 shows that hayrake tines are usually made of alloy steel. This means that broken tines can be used to make excellent centre punches.

Anneal the broken tine by heating to cherry red, cooling slowly. When soft, saw it to the desired length and grind the point to an angle of about 90 degrees. Next, fully harden the lower part by heating to cherry red and quenching in water. The procedure is fairly simple but there are one or two mistakes to avoid. Ensure that the cutting edge is not overheated if the flame is allowed to play directly on the cutting point. Keep the flame well below the cutting edge and allow the point to be heated by conduction. Quench after it has reached the cherry red colour.

Do not harden more of the tool than necessary. The lower end must now be cleaned with emery paper so that the correct tempering colour can be observed. Then carefully re-heat the lower end until the point is a brown to purple colour and requench.

Hayrake tines also make excellent pin punches and small chisels. For chisels grind the cutting edge to about 50 to 60 degrees for normal use. For punches and chisels leave the head soft so that fragments of metal will not fly off when it is struck. Also chamfer the head to slow down the burring caused by hammering. Other suggestions for reusing scrap materials are:

- Worn-out files can be turned into metal scrapers. Tempered to a straw colour.
- Flat springs welded to steel plough shares can double or treble share life.
- Worn-out plough disc coulters cut in half and welded to a handle can be used for lawn edge trimmers.
- Tractor track rod ends can be used for front end loader pins.

But before re-using a component for another purpose, remember that its strength depends upon not only what it is made of but also how it was heat treated. It is not advisable to use 'home-made' drawbar pins unless advice has been obtained from an engineer who has knowledge of the material concerned.

When using an oxyacetylene torch to heat machine parts to help to dismantle or assemble them — for example, warming bearings prior to fitting without the use of an oil bath — remember that you may be destroying the original heat treatment.

NUTS AND BOLTS

Only five of the many types of screw threads are commonly used in farm machinery. These are British Standard Whitworth (BSW); British Standard Fine (BSF); British Standard Pipe (BSP); Unified Course (UNC); Unified Fine (UNF); and the latest to be introduced, ISO metric. But the terms used to describe threads are common to all.

The pitch is the distance between a point on one thread and an identical point on an adjacent thread measured parallel to the thread axis. It is also the distance moved by a nut in one complete revolution (Fig. 1.52).

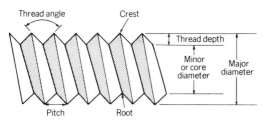

Fig. 1.52 Parts of a thread.

The thread angle is the angle formed between the inclined faces of the thread.

The minor diameter, also known as the core or root diameter, is the diameter of the threads measured across their roots at $90°$ to the axis.

The major diameter, also known as the nominal diameter, is the diameter of the threads measured across their crests or tops at $90°$ to the axis.

As for the screws themselves, BSW is a coarse thread with an angle of $55°$ and many examples of this can still be found on machines in use today. Its main disadvantage is that components may work loose because of vibration. Some form of locking device should be used with it.

BSF has been used in the past in machines where shock and vibration are common, such as on vehicles. The BSF thread form is the same as the BSW, but it has many more threads to the inch.

BSP threads also have the same form as the Whitworth, but there are many more threads to the inch in relation to the diameter, so that thin-walled pipe can be threaded.

The nominal size of a pipe refers to the bore size and not the external diameter. Only in metric copper pipe and some tubing does the stated size refer to the external diameter.

During the past 30 years the UNC thread has gradually replaced the Whitworth in vehicles and machinery. It has the same number of threads to the inch as Whitworth on all sizes, except the half-inch, where the UNC has 13 threads to an inch compared with 12 on the Whitworth. For most applications UNC and BSW screws and bolts are interchangeable despite the UNC having a 60 degree thread angle and a slightly different thread form.

UNF is the fine version of UNC, but is not interchangeable with its older counterpart, BSF, because the UNF has more threads to the inch.

Older reference books may quote the letters NC and NF. These refer not to the Unified thread system, but to the American National Course (ANC) and American National Fine (ANF), a thread system used in the USA before the war. This thread is known as the American National SAE or Sellars thread.

ANC and ANF bolts are, for most practical purposes, interchangeable with UNC and UNF bolts, except the 1 in ANF, which has more threads to the inch than the 1 in UNF—14 compared with 12. Since the war, the Unified thread system has been adopted in the USA, Canada and Britain.

The ISO metric thread is now being used on most agricultural equipment. A metric bolt is designated by its nominal size in mm and its pitch, which is also expressed in mm.

The ISO metric thread system is reasonably universal, but differs from the old SI metric system, upon which it is based. The practical significance of this is that an ISO metric nut will always fit an SI metric bolt, but the reverse does not always hold true.

There are various means of identifying screw threads, but in each case the nominal diameter and the number of threads to the inch need to be known. Both can be found with the aid of a ruler. However, the job is made much easier by using a thread gauge (Fig. 1.53).

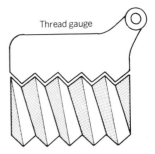

Fig. 1.53 Thread gauge.

Table 1.2 Thread table

BSW				ISO metric		
Diameter in	Threads per inch	Tapping size in		Diameter mm	Pitch mm	Tapping size mm
¼	20	³⁄₁₆		6	1	5
⁵⁄₁₆	18	¼		8	1.25	6.7
³⁄₈	16	¹⁹⁄₆₄		10	1.5	8.4

Having established the tpi and the diameter, refer to thread tables (table 1.2). For example, a bolt with a diameter of ⁵⁄₁₆ and 18 threads an inch could be ⁵⁄₁₆ Whitworth (BSW). A bolt with a diameter of 8 mm and a pitch of 1.25 mm, is an M8 x 1.25 (table 1.2).

It is not easy to distinguish between UNC and Whitworth (BSW) threads, as they have the same number of threads to the inch on all diameters except ½, (UNC 13 tpi; Whitworth 12 tpi). Try a spanner on the head of the bolt, because Unified, BSW-BSF, and ISO metric bolts each has its own head size. Many bolts and nuts carry a mark indicating their thread type.

Unified nuts are often marked with connected circles on one flat side or a circular groove turned on one face. Unified wheel nuts are sometimes marked by notches in the corners of the flat sides. These notches are sometimes used to denote a left hand thread.

Unified bolts and screws often have a circular depression in the head, or connected circles on one side, or, they may be marked UNC or UNF. The ISOM mark or the letter M on a bolt head indicates that the bolt has an ISO metric thread as in Fig. 1.54.

Replacement bolts must be made of the correct material, especially shear bolts. Several methods of coding a bolt are used to indicate the tensile strength — how much pull or stretch it will stand.

A bolt with a tensile strength of 50 tons per sq in will have a head carrying the letter S, or the SAE grade 5 marking (three radial lines). If it is an ISO metric bolt it may well be coded 8.8. The ISO metric code is arranged so that multiplying the two figures on the head together gives the yield stress in kg/sq mm. This is the stress required to stretch, and therefore weaken, the bolt. In the case of a bolt coded 8.8 the yield stress is 64 kg/sq mm.

Fig. 1.54 Markings on unified and metric fasteners.

Nuts are often marked with a code in the
form of a clock face (Fig. 1.55). The single dot
indicates 12 o'clock. The second mark a bar,
indicates the grade. In the case of a grade 8 nut,
the bar will be at the 8 o'clock position. *Note*:
the above markings are not always adhered to
by some manufacturers of fasteners. Also some
of the metric strength coding marks may appear
on fasteners carrying a unified thread.

Fig. 1.55 Markings on a nut.

Order a replacement bolt armed with as
much of the following information as possible:
Overall length and nominal diameter; length
and type of thread; grade; and type of head —
whether it is hexagonal, square, countersunk,
slotted, etc.

CHAPTER 2
Electricity Explained

WHAT IS IT?

All materials are composed of atoms which in turn consist of many smaller particles. Some of these smaller particles are known as *electrons*. In certain substances the electrons can move fairly easily from atom to atom. A stream of electrons moving through a material by passing from atom to atom is called an electric *current* and it is measured in *amperes*.

In order to cause a current to flow a driving force is required. *Voltage* is the name given to the force which tries to drive a current through a substance. A voltage can be provided by a battery or a generator. It is important to realise that a voltage can exist when there is no current flowing in a circuit.

Materials which readily allow a current to flow through them are known as *conductors*. All metals are good conductors. The human body is also a reasonable conductor and a fairly low voltage can push through a large enough current to kill.

All substances, including conductors, offer some opposition to the passage of electricity and this is known as the *resistance* and measured in *ohms*.

Thick cables offer less resistance than thin ones. Battery and welder cables are always large because they have to carry very large currents.

Substances like rubber and plastics have such a high resistance that for all practical purposes no electricity can flow through them. Such substances are known as *insulators*.

Electrical power is measured in *watts* (W) or kilowatts (kW). The wattage of a particular load is calculated by multiplying the voltage across it by the current flowing, thus $W = VI$ (I is symbol for current). For example, a motor operating from a 12 V supply and drawing a current of 20 amperes will use 240 watts or 0.24 kilowatts. To summarise so far:

$$\text{volts} = \text{amps} \times \text{ohms}$$
$$or \quad \text{ohms} = \frac{\text{volts}}{\text{amps}}$$
$$or \quad \text{amps} = \frac{\text{volts}}{\text{ohms}}$$

also

$$\text{watts} = \text{volts} \times \text{amps}$$
$$or \quad \text{amps} = \frac{\text{watts}}{\text{volts}}$$
$$or \quad \text{volts} = \frac{\text{watts}}{\text{amps}}$$

Electric currents

The type of current supplied by a battery or from a tractor generator is called a *direct current* (d.c.). This means that it will always flow in the same direction through the circuit provided the connections to the supply are not interchanged.

An *alternating current* (a.c.) such as provided by the CEGB in the UK is one which flows first one way through a circuit, then fades momentarily to zero and then flows the opposite way through the circuit. The UK power supply completes fifty such cycles every second and is therefore said to have a frequency of 50 cycles per second or 50 *hertz* (Hz).

The circuit

Electricity will only flow in a conductor which is a closed loop (or *circuit*). There are two common types of circuit, *series* and *parallel*. Figure 2.1(a) shows a series circuit. The same current must flow through each component and if any one part of the circuit should become broken then no current will flow and the entire circuit will fail. Figure 2.1(b) shows a parallel circuit.

Note that in a parallel circuit the current will be divided between the various loads. The branch of the circuit with least resistance will obviously draw most current. Tractor head lights are normally connected in parallel, so that in the event of one light failing the other will continue to operate.

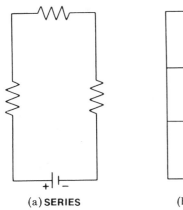
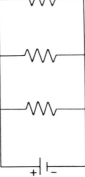

(a) SERIES   (b) PARALLEL

Fig. 2.1 Two types of electrical circuit.

Most circuits incorporate a special weak spot which is designed to melt if the circuit is overloaded. These weak spots are called fuses. Figure 2.2 shows a conventional fuse but many circuits are also protected by circuit breakers and fuseable links. These are short lengths of wire which melt in the same way as a fuse in the event of a major overload.

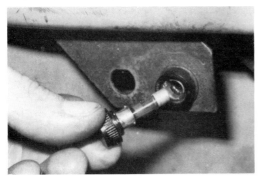

Fig. 2.2 Typical fuse.

Ammeters and voltmeters

Small instruments such as shown in Fig. 2.3 are fairly cheap. These can be used to measure voltage, current and resistance. *Note:* A voltmeter must be placed *across* a component or the supply, *not* in the circuit.

An ammeter should be placed in the circuit and measures the current flowing through the instrument (see Fig. 2.4).

Never confuse the use of the two instruments. Obviously an ammeter must have a low resistance in order to carry the circuit current.

Therefore if this instrument is placed across the supply it will draw an excessive current and burn out almost instantly.

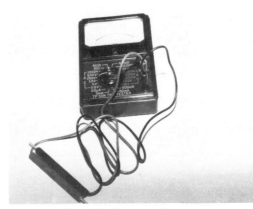

Fig. 2.3 Cheap multimeter.

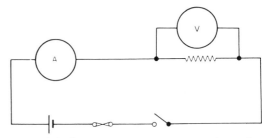

Fig. 2.4 Correct way to use an ammeter and voltmeter.

THE BATTERY

The battery is a good servant, but a poor master. Handle it with care — it will last longer and be less dangerous.

Common mistakes

1. *Overfilling:* This causes the electrotyte, which is diluted sulphuric acid, to bubble out of the battery top. Spilt acid causes corrosion of the terminals and the battery housing which can prevent the current getting to the starter.

2. *Discharged batteries:* Left for prolonged periods, lead-acid batteries will gradually discharge themselves, even when they are in perfect condition. The higher the temperature, the faster the discharge. Store the charged battery in a cool place. But remember:

(a) The electrolyte in a discharged battery is mainly water and it can freeze. Combine batteries are often found to be cracked in the spring when they have been left on the machine all winter.

(b) The plates of a battery left discharged too long develop a hard lead sulphate coat which reduces the normal chemical action. Sometimes a 'sulphated' battery can be restored by charging it slowly over a long period.

3. *Overcharging:* Water is lost from the cells because overcharging splits the water into hydrogen and oxygen which bubble through the battery and tend to wash the active material from the plates. When a battery uses a lot of water check the charging system. The slower a battery is charged the better. To test the charge, take three consecutive hydrometer readings (see table 2.1) at hourly intervals. When there is no rise in specific gravity, the battery is charged. Badly sulphated batteries may take 60 to 100 hours to recharge.

Table 2.1

Battery condition	Specific Gravity
Charged	1.28
Half charged	1.20
Discharged	1.11

4. *Corrosion:* Clean the battery regularly or the acid fumes during charging will combine with dust and dirt to form a film and short-circuit across the top. This can discharge the battery quite rapidly when the tractor is idle. As there is a limit to the number of times a battery can be charged and discharged, this will shorten its life (Fig. 2.5).

Fig. 2.5 Corroded terminal.

5. *Loose clamp:* Vibrations in work can crack or wear a hole in the case or damage the battery internally when the clamp is loose. When the clamp has corroded away, fix another.

6. *Poor connection:* Replace loose or corroded terminals with the clamp type, such as shown in Fig. 2.6. These are attached by screws and do not need to be soldered. Melt the old terminal off the cable with the gas torch.

Fig. 2.6 Replacement terminal.

Testing

First charge the battery, and leave it for a few days so that it can discharge itself. Operate the starter motor for 20 to 25 seconds with the tractor stop control out. Wait a few minutes for the starter motor to cool. Operate the starter for another 15 to 20 seconds. When it passes this test, look for trouble in the charging system. Otherwise, check for leaks, loose terminal posts and general neglect.

The hydrometer can be used to measure the specific gravity of the electrolyte. The heavier the electrolyte the more charge in the battery. Such readings only apply when carried out at 20 to 27°C. Every decrease of 15°C in temperature increases the specific gravity by 0.01. The hydrometer is cheap and easy to use, but unless a battery is known to be in good condition, it gives little indication of the battery's ability to deliver current (Fig. 2.7).

The high-rate discharge tester will test the battery under loaded conditions. Test each cell in turn by placing the connecting prongs across the cell-joining links for 10 to 15 seconds and noting the cell voltage. A good cell should sustain a voltage of 1.3 to 1.5 volts.

Voltage falls rapidly on faulty cells. Never use this tester on a battery suspected of being less than 75% charged.

When the battery is on a tractor which has just been running, operate the starter for a few seconds before testing to remove the 'surface charge'.

Batteries with no exposed cell connectors can be tested with a special high-rate discharge tester which fits across the battery terminals and tests the whole battery at once.

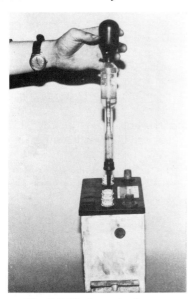

Fig. 2.7 Hydrometer in use.

Battery rules

● Always disconnect the battery earth lead first when removing the battery and when replacing it, refit the earth lead last. This avoids the possibility of a short-circuit between the live terminal and earth.
● Charge slowly.
● Top up with distilled water or use clean ice from a plastic-lined refrigerator after defrosting.
● Trickle-charge when not in use.
● Remove any spilt acid from clothing or skin immediately and seek medical help when acid has splashed into the eyes. As a first-aid measure, wash the eyes in running water for 10 to 15 minutes while waiting for medical attention. Acid spilt on clothes may be neutralised by a solution of ½ kg of baking soda in five litres of water.
Never:
● Smoke or hold a naked light near a battery which is or has recently been on charge. Gases from a battery on charge are highly explosive.
● Leave spanners on the battery.
● Overcharge.
● Allow a battery to remain discharged.
● Overfill.
● Leave in a dirty or corroded state.
● Scrap a battery on hydrometer evidence. Check with a high-rate discharger first.

Battery carrier

This simple battery carrier is made from 'Meccano' type angle iron, with a plywood floor and a small pair of wheels bolted to the frame. A length of 25 mm diameter pipe makes a good handle with which to push it around. Use it when you have to move batteries around the workshop, or for taking batteries out of tractors on the farm (Fig. 2.8).

Fig. 2.8 Battery carrier.

TRACTOR DYNAMOS

Dynamos on tractors rarely give trouble. If they do break down a service exchange unit costs only a few pounds.

The main problem is brush wear. Brushes are easily replaced — a new pair costs a few pence — if caught in time, and it is worthwhile inspecting them every few months to check for wear. Look through the holes in the back plate of the dynamo and if the brushes are less than 5 or 6 mm long they need renewing. Otherwise they will wear until the wire connected to the brushes comes into contact with the armature and wears a deep groove in the copper segments of the commutator as it revolves.

First step in dismantling a dynamo is to remove the two long retaining screws in the back-plate. Lift the back-plate, complete with brushes and springs, gently off the main dynamo body (Fig. 2.9).

Undo the screws holding the brush wire and remove the brushes from their guides. Lift the springs off their locating pins and wash the back-plate in petrol to remove dirt and grease.

While the back-plate dries remove the main dynamo case and inspect the commutator. If there is a ridge more than 1.5 mm deep where the brushes have been in contact with it the commutator will need skimming down on a lathe — a job for the dealer. Light rubbing with glass-paper will usually remove any carbon smear and small surface pits and dents.

Fig. 2.9 Remove the end-plate, with brushes, first.

After cleaning up the commutator the mica insulation must be undercut and removed below the level of the commutator surface. Do this with a thin knife blade or small hacksaw blade, pulling the blade away from the arma-ture. Blow off with an air-line to remove loose particles, which should be wiped clean from the rest of the dynamo.

Reassemble and fit new brushes, making sure they slide freely in their guides and screws are tight. If the springs are broken replace them. Fit the end-plate, pushing the brushes over the commutator with a small screwdriver and taking care not to chip the brittle carbon. Ensure the endplate dowel is properly located in its recess. Make sure the brush wires do not get trapped between the end-plate and main body of the dynamo. Tighten the two end screws and re-fit the dynamo to the tractor. The whole job will take about an hour and will make a dynamo fit for many more hours of service (see pictures in Fig. 2.10).

Testing the dynamo

To check if a dynamo is working, remove both leads (from the D and F terminals) and join the two terminals together with a short wire. Then connect a voltmeter between the two joined

A new dynamo is the only answer in this case.

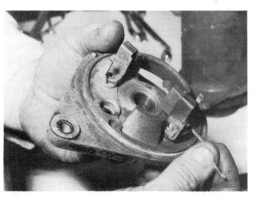

Make sure the brushes slide easily in the guides.

Rest the armature on a vice while undercutting the copper segments. Remove only a few thousandths of insulation.

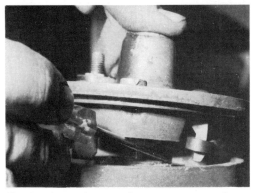

Gently prise brushes over the commutator to refit.

Fig. 2.10 Repairing a dynamo.

terminals and earth. Run the engine at about 1000 r.p.m. (in the case of a tractor) and note the voltage. At 1000 r.p.m. the voltage should be at least 20 V. If the dynamo passes this test, repeat the test at the control box, but this time remove the wires from the D and F terminals and join the *wires* together. No voltage here indicates faulty wires. Adequate voltage indicates a faulty control box.

ALTERNATOR CHECKS

Odd as it may sound, an alternator does in fact produce a direct current like a dynamo. This is because the alternating current is rectified to a direct current by an electronic device known as a rectifier which is housed in the back of the machine. Alternator trouble is generally indicated by a discharged battery. Usually the charging light will also be illuminated to confirm that no charge is being produced.

Before an alternator is condemned various checks should be made to ensure that it is not replaced unnecessarily. These checks are described below using a Ford tractor fitted with a Lucas ACR alternator.

When the warning light is not operating at all, check its circuit as follows: first remove the plastic plug from the back of the alternator and turn on the key start switch. Then short the *small* connection in the plug to earth, as shown in Fig. 2.11 (the exhaust manifold makes a convenient earthing point). Take care not to short the large brown output cable connection, as this is live. When the small connection is earthed, the charging warning light should be illuminated. If it does not light, check the bulb and its circuit. If the bulb does light, the problem lies within the alternator itself. *Note:* some alternators will not generate electricity unless the warning light circuit is complete because this circuit supplies the initial current needed to start the charging process.

Usually, alternator trouble results in the warning light staying on when the engine is running. When this happens, first investigate all wires and connections. If the tractor has a battery temperature sensor, (Fig. 2.12) check its circuit by removing the sensor lead from the *alternator* terminal marked S and connecting a voltmeter between this lead and a good earth. When the key start switch is turned on, the battery voltage should be recorded. If battery voltage is not recorded, the battery temperature sensor circuit should be thoroughly checked and if this proves to be intact the temperature sensor should be replaced or tested by substituting a unit from another tractor. Strictly speaking, the output current of an alternator should be checked by temporarily fitting an ammeter in the output lead, as it is possible for some faults to cause the warning light to glow when the alternator is still producing

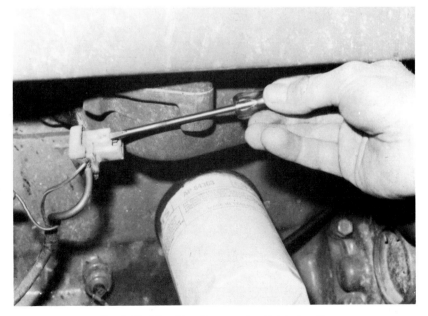

Fig. 2.11 Checking the warning light circuit.

Fig. 2.12 Battery temperature sensor.

current.

Turning to the alternator itself, having removed the back cover, there are three components which are very simple to test and replace. First, check the voltage regulator by shorting it out and running the engine at about 1000 r.p.m. and noting if the warning light is extinguished. With the type of regulator shown, this is achieved by shorting its casing to earth (Fig. 2.13). If shorting it out causes the alternator to start charging, then a new regulator should be fitted. This will take up to one hour in time and will cost several pounds. A new regulator may only have two wires instead and may well be a different type, such as shown in Fig. 2.14.

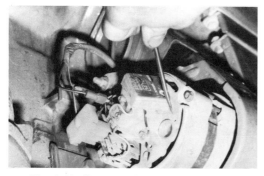

Fig. 2.13 Shorting the voltage regulator.

Fig. 2.14 Replacement voltage regulator having two leads.

The next test is to remove the surge protection diode and test it. Figure 2.15 shows a typical surge protection diode. This should only conduct current in one direction and can be tested by wiring it in series with a 12 V battery and a 2.2 W test lamp (as shown in Fig. 2.16) and noting if the bulb lights. Then reverse the connection on the battery and note if the bulb lights. The bulb should only light on one test. If it lights both times, the diode is faulty. A new one will take about ten minutes to fit and will cost only a few pounds.

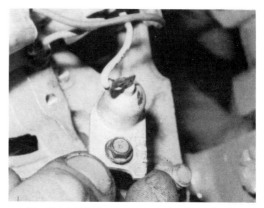

Fig. 2.15 Typical surge protection diode.

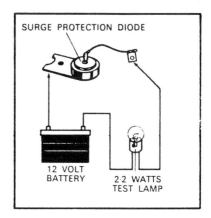

Fig. 2.16 Testing the surge protection diode.

Thirdly, remove the brush mounting box and replace the brushes if these are shorter than 8 mm (see Fig. 2.17). Also remove the carbon dust from the brush box. Brushes will take about one hour to fit and are inexpensive to buy.

Figure 2.18 shows a typical Lucas rectifier pack. This can be removed and tested in a similar manner to the surge protection diode, but consult the relevant workshop manual first.

Fig. 2.17 Inspecting the brushes.

Fig. 2.18 Typical rectifier unit.

To prolong the life of an alternator, NEVER:

(1) *Reverse the battery connections.* This will cause instant damage to the electronic components in the alternator.
(2) *Make or break any part of the charging system when the engine is running.* This includes removing the battery.
(3) *Short the output lead to earth with the engine running.*
(4) *Steam clean the alternator.* The heat will also damage the electronic components.
(5) *Arc weld on the tractor* or an attached machine, without first disconnecting the alternator leads.

STARTING SYSTEMS

Figure 2.19 shows the layout of a typical pre-engaged starter system.

When the starter key is turned, a current flows from the solenoid live terminal, through the key start switch, through the safety switch and back to the solenoid operating terminal. (The safety switch ensures that the tractor will not jump forwards or backwards when the key is turned, and is usually 'cancelled' by placing one of the gear levers in neutral.)

The solenoid is now magnetic and the plunger is pulled towards the centre of the solenoid, pulling with it a lever which moves the starter pinion into mesh with the flywheel ring gear. The final part of the plunger's travel closes the main contacts in the solenoid, which

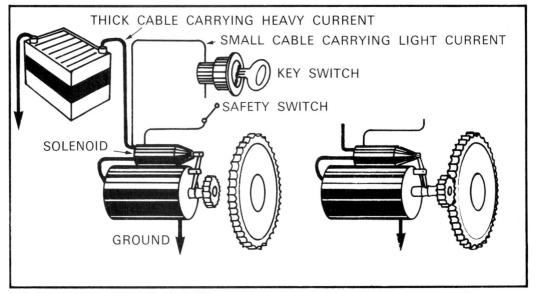

Fig. 2.19 Layout of pre-engaged starter system.

connect the battery to the starter. A heavy current can then flow from the battery, through the solenoid contacts and to the starter input terminal.

When a starter refuses to operate satisfactorily with a well-charged battery, switch on the lights and watch what happens when the starter key is turned. If the lights dim but the motor does not turn, the most likely fault is in a connection, probably the battery terminals or the power input connection on the solenoid. But the link between the solenoid and the starter may have shorted to earth or there could be a short circuit inside the starter motor.

Should the connections appear in order, the starter motor is probably at fault. Even if any of the above connections are causing trouble, the solenoid should still be heard to click.

Should the lights not dim when the key is turned and the starter does not crank the engine, the starter circuit should be investigated, using a 12 V light bulb to pin-point the trouble.

No sound at all when the key is turned, points to a fault in the solenoid circuit. Check

Fig. 2.20 Checking the small wire has not dropped off the solenoid.

Fig. 2.21 Check the safety switch connections.

that the wire has not dropped off the solenoid or the safety switch. (See Figs 2.20 and 2.21.) A quick check to find whether the starter motor and solenoid are operating is to put the tractor in neutral, then join the solenoid live terminal to the solenoid operating terminal (see Fig. 2.22). If the starter now operates, the trouble must lie with a faulty safety switch or key start switch or a break in the cables joining these components to the rest of the circuit.

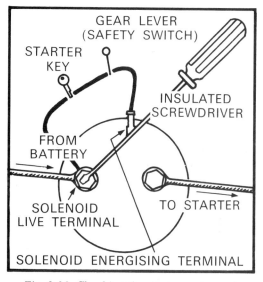

Fig. 2.22 Checking the starter motor and solenoid.

Some solenoids are earthed through the brushes in the starter and it is possible for certain starter faults to render the solenoid inoperative.

A rapid clicking noise from the solenoid when the key is turned indicates that either the battery is in a poor state of charge or the solenoid has a faulty coil.

When the starter motor cranks the engine too slowly, the battery is usually to blame. Try another battery because the starter motor can produce the same symptoms because of worn or sticking brushes, a commutator which is shorted out with dust from the brushes or, possibly, fuel oil which may have been spilt on the motor.

A starter motor cranking the engine too slowly or not at all will have to be removed for inspection. Disconnect the battery earth lead and make a note of the positions of the various cable connections.

Table 2.2 Common starting faults

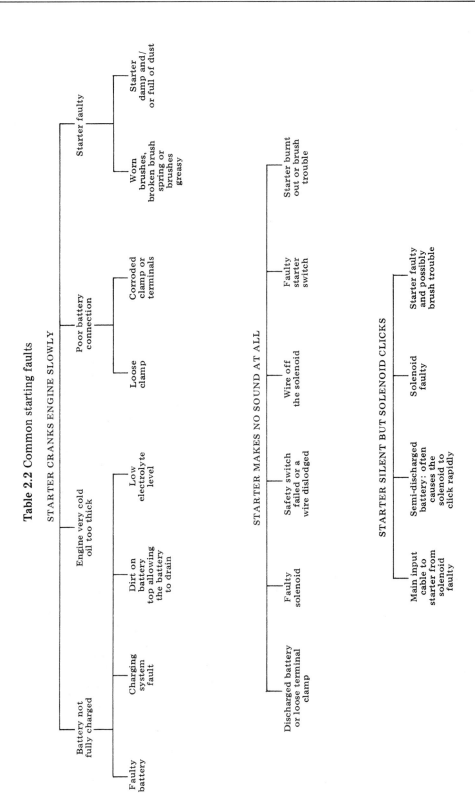

STARTER CRANKS ENGINE SLOWLY

Battery not fully charged
- Charging system fault
- Faulty battery

Engine very cold oil too thick
- Dirt on battery top allowing the battery to drain
- Low electrolyte level

Poor battery connection
- Loose clamp
- Corroded clamp or terminals

Starter faulty
- Worn brushes, broken brush spring or brushes greasy
- Starter damp and/or full of dust

STARTER MAKES NO SOUND AT ALL

Discharged battery or loose terminal clamp
- Faulty solenoid
- Safety switch failed or a wire dislodged

Wire off the solenoid
- Faulty starter switch
- Starter burnt out or brush trouble

STARTER SILENT BUT SOLENOID CLICKS

Main input cable to starter from solenoid faulty
- Semi-discharged battery: often causes the solenoid to click rapidly

Solenoid faulty
- Starter faulty and possibly brush trouble

Fault finding

Table 2.2 is a summary of common starting faults and their symptoms, all of which can easily be traced by a farm mechanic without the use of specialised equipment. Many minor faults can be rectified in minutes.

TRACTOR LIGHTS

Lights on a tractor being used on the public highway must, by law, be maintained in working condition. It is possible to be prosecuted for driving a tractor with defective lights on a public road in daylight.

A tractor with headlights and no sidelights could be mistaken for a much narrower vehicle. Over-hanging implements must also be lit.

The only equipment needed to test the lighting circuit is a test lamp.

Tractor-lighting troubles are usually due to vibration, metal corrosion and deterioration of the insulating materials. The most common fault is a loose connection in the wiring harness or wire connections which have been pulled apart.

When both headlamps fail simultaneously, it is probably due to the wiring rather than the lamps. A typical trouble spot (Fig. 2.23) is where the feed-wire branches into a left and right supply. Joins made with bullet connections are prone to corrosion. Such a fault can be traced by switching on the lights, noting if they light up and carefully wiggling each connection in turn.

Fig. 2.23 Typical trouble spot.

Any loose or dirty connection should be pulled apart, cleaned and pressed firmly back together. These connections have a tendency to corrode so badly that the wire pulls out altogether, in which case a new bullet end must be fitted.

The lighting circuit can fuse. Normally each

lamp is on a separate fuse to prevent a black-out. Fuses are usually easy to get at (Fig. 2.24), but a blown fuse does not necessarily indicate a fault. It can age and melt when carrying its normal current.

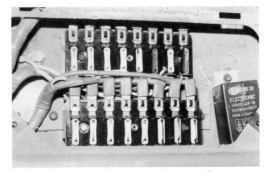

Fig. 2.24 Multiple fuse-panel.

Often, a blown fuse can be spotted because the small wire or metallic strip carrying the current is no longer intact.

A small break indicates a small overload. But when the wire appears to be blown to pieces a more severe overload or even a short-circuit is probable. A short-circuit is often caused when a live wire touches earth and has been the cause of many fires.

When all the lights fail simultaneously, the trouble could be a faulty lighting switch or the connection may have been pulled off the switch.

To remove a bulb or sealed-beam unit, undo the screws which secure the front of the lamp and remove the bulb-holder and bulb from the rear. Sealed-beam units are fastened to the wires by special holders (Fig. 2.25) which press on to prongs on the back of the sealed beam. The sealed-beam unit is usually

Fig. 2.25 Removing a sealed beam unit.

located with a small notch so that it is fitted the correct way up. When the unit is mounted in rubber take extra care to locate the notch correctly.

A bulb can usually be inspected by eye. A good bulb should be clear and the filament wires intact. A blackened bulb will probably be defective, while a bulb with a broken filament wire will definitely not work.

A better method of testing a bulb (Fig. 2.26) is to connect it across a battery of the same voltage or fit it in a lamp known to be working and see whether it lights up.

Fig. 2.26 Testing a bulb.

Should the bulb work, check that the contacts in the lamps are clean and spring-loaded to press against the bulb (Fig. 2.27). When the contacts are firm, check that the current is reaching one of them by connecting a test lamp between each contact and a good earth.

Fig. 2.27 Checking a bulb-holder spring.

Should the test bulb light up, the trouble is almost certainly a poor earth. The earth connection is made either through the body of the lamp or, more often, a separate earth wire (black) is fitted to the bulb holder and is

connected to the tractor body at a convenient point (Fig. 2.28).

Fig. 2.28 The earth connection.

A poor earth connection can cause the lights to dim or glow faintly.

Tractor brake lights are usually operated by the brake pedals working a small switch (Fig. 2.29) which is prone to stick in the 'off' position.

When both brake lights fail, check the brake switch. Disconnect the two wires and join them together. If the switch is the cause of the trouble, the lights should now work. Remember that some brake lights will not work unless the starter key is turned to its first position.

Fig. 2.29 Brake light switch.

When replacing a bulb, fit only the correct wattage. Too low a wattage will result in a dim light and too high wattage will create too bright a light. Excess heat can also damage the lamp. The extra current demand of the bulb could overheat the feed wires when the circuit is not fused. An incorrect bulb in the direction indicators will change the speed of flashing.

ENGINE PRE-HEATERS
Glow plugs

A glow plug system consists of a glow plug in each pre-combustion chamber, a resistor to reduce the current flow and a heavy-duty switch connected in series so that the current flows through each plug in order to complete its circuit.

Should one plug or other single component fail, the circuit is broken and no current flows to any plugs. Often the current reducing resistor is visible to act as a temperature indicator by glowing red when the plugs have also reached the correct temperature. Glow plugs are normally operated by switching on for 20 to 30 seconds or until the indicator glows red.

Plug failure because of its heating element burning out is a common fault (Fig. 2.30). To find the ailing plug, short out each plug in turn by joining the input and output leads with a thick cable and checking whether the system then operates. Once the offending plug has been located, the shorting cable can be left in position to start the engine in an emergency.

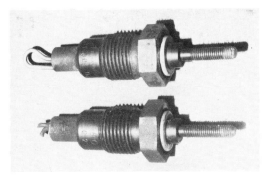

Fig. 2.30 Failed and sound glow plugs.

Never operate the system more than once with one plug shorted out, as the remaining plugs will be overloaded. Most glow plug manufacturers advise against using the system even once with one plug bridged.

When this method fails to locate the trouble, inspect all the wiring, paying extra attention to the current-limiting resistor (Fig. 2.31). Always replace where necessary with the correct type. Each of several resistors available has a different resistance and fitting the wrong one could cause an excessive current to flow and destroy other plugs in the circuit. Figure 2.32 shows two different types of resistor.

To check whether electricity is reaching the current reducing resistance, connect a 12 V test lamp between its live terminal and earth. Further checks require a voltmeter, otherwise remove all the plugs and inspect them.

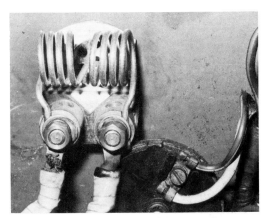

Fig. 2.31 Current-reducing resistor.

Fig. 2.32 Two types of resistor.

When the current reducing resistor appears to heat up too quickly, check that it is the correct one, then look for a glow plug or wire which is shorted to earth. A plug shorted to earth will cause the resistor and remaining plugs on the live side of the short to be overloaded while the plugs on the earth side receive no current.

Frequent failure of glow plugs could be caused by the system being left switched on once the engine is running. Heat and shock waves of burning fuel will soon destroy a hot glow plug. Glow plugs seldom fail through being defective. Nearly always they are destroyed by other factors such as faulty injectors, injection pressures which are too low, or incorrect injection timing.

Thermostarts

The thermostart is fitted in the inlet manifold where it heats the air that is drawn into the engine. It consists of a heater coil, an igniter coil and a valve arrangement which opens to allow fuel oil to run on to the igniter coil once it is hot (Fig. 2.33).

Fig. 2.33 Thermostart removed from the manifold inlet.

The fuel is usually supplied by gravity from a small reservoir topped up from the fuel leak-off system. Sometimes the leak-off pipe acts as a reservoir. The thermostart is normally operated for about 15 to 20 seconds before the engine is cranked.

To test a thermostart, check that current is reaching it with a test lamp between its cable and a good earth and switching on. If the power is in order, remove the unit from the manifold, reconnect its fuel and electrical supplies, and make sure the unit is making good electrical contact with the tractor body.

Use a separate wire to make the earth connection if necessary (Fig. 2.34). Switch on and see whether its coils start to glow red. Eventually a small flame should appear. Have an assistant standing by with a fire extinguisher — just in case.

Fig. 2.34 Separate wire ensures a good earth during testing.

Thermostarts can be 12 V or 24 V, so check that a replacement is the correct one. Do not cross thread the unit in the manifold, which will probably be made of an aluminium alloy.

On some systems, check that the fuel reservoir contains adequate fuel before operating

the thermostart for the first time after installation, otherwise the unit may not work and could be damaged.

CHAPTER 3

Workbench Wisdom

Today, when profit margins can be wiped out merely by failure to get a broken machine quickly back into action, the man whose farm 'workshop' is merely an old 'ammo' box of assorted tools is putting himself and his farming at hazard. Even if facilities consist of no more than a permanent workbench which provides an ordered place for tools and accessories there is already some guarantee against frustration and despair — as will be testified by anyone who has torn his knuckles and mangled components when trying to repair a machine with the wrong tools because the right ones cannot be found.

The bench is the starting point and whether it is a highly sophisticated, bought-in affair or home-made from odd bits of timber, it should be solid and roomy. A metal top is to be preferred and is not such an extravagance as it sounds; wooden ones can become oil sodden and a metal sheet preserves a smooth, hard-wearing working surface.

Site the bench under a window if possible and in any case make sure it has adequate artificial light. Racks full of spring clips or trays, divided shelves or other systems to keep things together, should be close by. A solid old sideboard picked up from a farm sale can serve as a crude but useful workbench with a sheet of metal added; old household chests of drawers can make splendid tool cabinets.

Simple wooden racks holding oil tins cut in half provide cheap but efficient storage for spares. The grooves in strips of corrugated sheet make ideal compartments for keeping small components apart; the axle and discs from an old set of disc harrows with a tripod welded on one end allows it to stand upright and form a 'cakestand' type of receptacle for nuts and bolts. These are a few of the ingenious ideas that have served well in farm workshops and have cost virtually nothing. We shall be discussing them in detail later. The main aim should be to create order out of chaos and keep it that way.

The workbench must have a vice — two if finances will stretch to it. A hefty one is needed for the big items and a smaller one for the more delicate parts. Secure the vice properly (See Fig. 3.1).

Fig. 3.1 The wrong type of bolt: the square shank in the round hole anchoring the vice gives little support and the vice will soon work loose. Over-tightening could result in the flange breaking off.

The bench must be big and strong enough to withstand rough treatment. The more working space available the better. Small benches soon become cluttered up with bits and pieces of machinery often to the point where there is no room for tools. It is no major operation to make a good large substantial work-top so the vice has the top of the jaws equal to the height of the worker's elbow from the ground, with the arm bent.

CHOOSING A VICE

It pays to have a vice that will stand up to such jobs as bending or riveting.

There are two main types, the leg-vice and the parallel-jaw vice. The leg vice has the reputation of being exceptionally strong. Blacksmiths prefer it; it is ideal for gripping metal which needs to be chipped, hammered or bent rather than the more delicate job.

Figure 3.2 shows its one big disadvantage; the jaws do not meet squarely due to the movable jaw being pivoted halfway down the support leg (see A in the drawing). This makes the jaw move through an arc and it is parallel to the fixed jaw only when the vice is closed. To hold work securely it is often necessary to grip it very tightly, which can result in distortion.

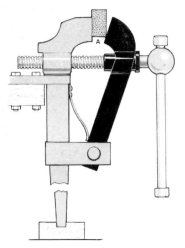

Fig. 3.2 A leg-vice. The leg may be let into the floor or supported on a wooden block firmly attached to the floor. It is more rigid than types fixed only to bench top, but is not so useful for engineering work.

The parallel-jaw vice was designed to overcome this problem and is now the most popular vice in engineering. Many different makes and types are available, but for all-round farm workshop a 12 cm model which opens to 15 cm, with renewable jaw-plates without quick release arrangement, will cope with the majority of work (Fig. 3.3).

A quick-release mechanism saves time but this type of vice does not stand up to hard work as well as the fixed-nut type. (Fig. 3.4).

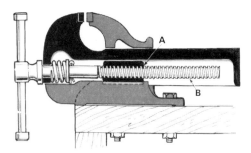

Fig. 3.3 A fixed nut parallel-jaw vice — the combination of the nut A being more or less part of the body and the square threaded screw B tends to make it much stronger than quick release vices.

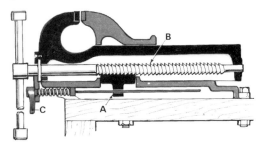

Fig. 3.4 Parallel-jaw vice with quick release arrangement. By pressing lever C the half nut A, which may be renewed when worn, is disengaged from the buttress threaded screw B.

It is worth paying a little extra for a swivel base vice. When dealing with large, awkward pieces of tackle or long lengths of pipe it is a great help to be able to position the vice so that the work does not foul a bench or wall. It may be swivelled in a complete circle and locked in any position.

The jaws of the vice are made from cast steel and serrated like a file. To prevent damage to soft metal or to bolt threads it is worth while making a set of protection pads for the jaws. Lead, hard wood or fibre are ideal materials.

Keep the vice in good working order. Check that the jaw-plates are tight and renew the set screws if they are worn. Clean the tightening screw threads occasionally to remove from filings and dirt. Care should be taken not to over-oil the threads.

TOOL KITS

There is a lot to be said for gradually building up a tool kit rather than buying it all at once. The contents depend largely on what equipment is to be maintained and how much do-it-yourself work is likely to be attempted. There are, however, a few basic essentials, such as spanners, hammers, pliers and screwdrivers, with spanners coming at the top of the list.

A golden rule in engineering is always to use the correct tool for the job in hand and this applies especially to spanners. Adjustable and Stilson wrenches are fine when used in the right application, but all too often small bolts and studs are sheared off because an adjustable spanner is used instead of a small ring one. Also there is nothing more annoying than coming across a metric nut in an awkward corner and finding that the only spanners available are Whitworth.

So before buying tools, take stock of the types of nuts used in your range of tackle. They might be BSW (British Standard Whitworth), Unified or Metric. Unified measurements are made across the nut and not across the bolt, which the nut is to fit. This measurement is known as AF (across the flats) and is referred to much more than Unified (Fig. 3.5).

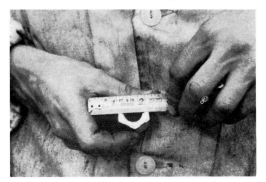

Fig. 3.5 Measuring across the flats of $1\frac{1}{8}$ in AF nut.

Open-ended spanners should be bought in complete sets rather than individually because it is always the one missing that is wanted in a hurry. Basic essentials should include BSW $\frac{1}{8}$ to $\frac{3}{4}$ in, AF $\frac{5}{16}$ to 1 in and metric 10—20 mm. Make sure the sizes overlap, for example, $\frac{1}{8}$ — $\frac{1}{4}$, $\frac{1}{4}$ — $\frac{5}{16}$, $\frac{5}{16}$ — $\frac{3}{8}$ and so on. Two spanners of the same size are often needed when releasing or retightening locking nuts.

SPANNER CARE

The jaws of a normal double open-ended spanner are set at an angle of $30°$ to the handle which is sufficient for most jobs, but there are, however, some which have jaws set at $75°$ or even $85°$. These types are called obstruction spanners and are used where it is impossible to get a $30°$ spanner on to a nut. Obstruction spanners are a luxury and would be seldom used in the farm workshop.

It is much better to buy a few double-ended ring spanners. They have 12 corners or points on the inner edge of the ring and can, therefore, be fitted to a nut in 12 different positions. A full set is unnecessary. Every alternate one in the range will be sufficient. Gaps can be filled by buying a box of sockets including both AF and metric sizes.

Spanners must not be abused as makeshift hammers and levers. Even though they are made from hardened tough steel, the jaws will eventually extend and start slipping off nuts. (Fig. 3.6). Fitting the spanner to a nut in the wrong way may also lead to damaged jaws. The correct method is with the smaller jaw arm towards the line of pull — which leaves the larger jaw to take the main strain (Fig. 3.7).

Always use the correct spanner for the size of nut or else carefully pack the jaw using a

Fig. 3.6 This method of increasing leverage may help to slacken a stubborn nut but it won't do the spanners any good.

washer or small plate (see Figs. 3.8 and 3.9).

The tool boxes fitted to tractors and implements are often so small that the farmer has to make up his own portable tool box. Carrying

Fig. 3.7 The quickest way to ruin an adjustable wrench: the small jaw is taking most of the strain.

the odd spanner and hammer is not enough for dealing with the more complicated tackle, such as combines and balers. A comprehensive selection of gear is required with the addition of a few bolts, nuts, washers, cotterpins and the odd part that is known to be suspect. An emergency supply of baler shear bolts can save a journey back to the workshop.

Tools from a portable tool kit will be used for ordinary work in the workshop, but a rule should be made that they are always replaced after use. To avoid searching for tools in the bottom of the box, fit it out with clips or divide it into compartments to hold the spanners and the smaller items.

SOCKET SPANNERS

Socket spanners, tubular and made of alloy steel, are usually 12-pointed each side allowing a nut to be moved 30°. The advantages

Fig. 3.8 Leave the adjustable wrench in the tool box when doing jobs like this: use the correct size open-ended or ring spanner.

Fig. 3.9 A washer makes a good packing piece when the only available spanner is too big.

are that there is little risk of damage to either nut or hands and access to a nut in a restricted position is easier. The spanners can also be used on square nuts without damaging the nut or slipping.

Sockets can be bought in metric sets but only the sockets themselves are in metric sizes. The fittings and drive bars (the piece that fits into the back of the socket) are still made in imperial sizes.

The driving bar can be ³/₈ in, ½ in or ¾ in square. If the spanners are to be used on heavy work, such as crawler track adjustment or undoing wheel nuts, buy a set with ¾ in drive bar. Half-inch drive will be sufficient for normal workshop use (Fig. 3.10).

Buy the best set you can afford; they are not cheap but they will last for years and the bigger the set the more accessories you get. Accessories in more expensive sets include the speed brace for tightening and undoing nuts quickly, a ratchet handle for the same purpose in restricted access, the universal joint for awkward nuts and bolts, and a set of long and short extension bars.

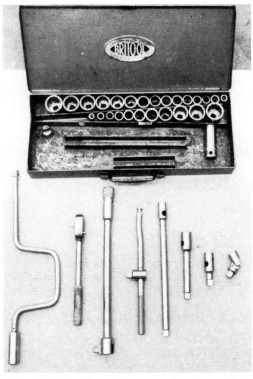

Fig. 3.10 Good quality comprehensive socket set.

Extra long sockets, for nuts and bolts in recesses or where a nut is on a long threaded bar, screwdrivers and sparking plug spanners can also be bought.

If you have many continental machines on the farm, and with more and more manufacturers turning to metric hardware, a metric set will be the best buy.

Maintenance is minimal; wipe down with an oily rag after use and put a drop of thin oil on the ratchet occasionally.

PLIERS

When selecting the smaller hand tool such as pliers it pays to shop around a bit to hunt out the types which have more than one use, such as a pair of square-nosed pliers with side wire cutters and insulated handles. Some incorporate a stronger cutting device either side of the hinge rivet and these are usually capable of handling wire up to 3 mm thick. They are the ideal tool for trimming and cutting split pins to length.

Here are a few of the many shapes and sizes of pliers available (Fig. 3.11).

Vice grips with wire cutters. It is best to buy a large pair of these. They can be used as pipe grips as well as for clamping metal for cutting, welding or drilling.

Multi-purpose pliers incorporating hammer head, wire cutters, staple remover, square and round gripping jaws and nail remover. An ideal tool for fencing jobs and in the workshop.

Square-nosed straight jaw pliers with insulated handles. The jaws are slightly tapered towards their ends and have small serrations to give a firm grip. Others are: round-nosed pliers, square-nosed pliers with straight and curved jaws, long-nosed pliers, pointed-nose pliers (handy for holding small nails or rivets in position), heavy duty long-nosed pliers with wire cutters, external circlip pliers with right angle points, and curved long-nosed pliers.

Special strong pliers are required for the snap-rings which are common in gear-boxes. Attempting to remove a snap ring without the correct pair of pliers will result in damage to both the tool and the ring. Figure 3.12 shows a strong pair of snap ring pliers being used to fit a ring.

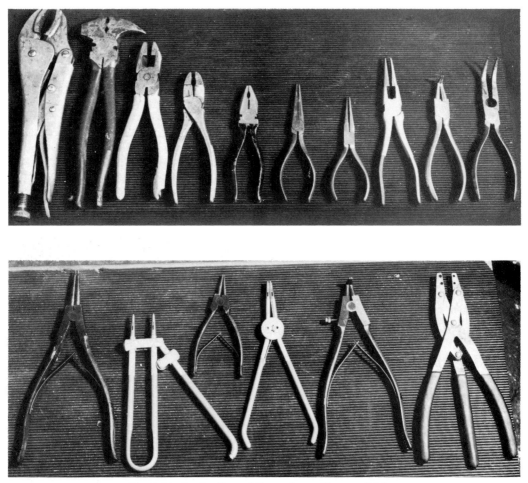

Fig. 3.11 A selection of the various pliers available.

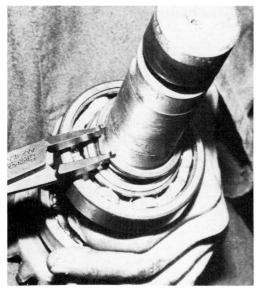

Fig. 3.12 A strong pair of snap ring pliers.

ALLEN KEY

A set of Allen keys is a useful accessory in a workshop. The keys are made of tool steel, are hexagonal in section, and have a right-angled bend in them. The bend allows either end of the key to be used while the other is gripped to exert twist (Fig. 3.13).

On farm machinery Allen keys are mainly used on small grub screws to hold bearings, lock-collars and baler knives in position (Fig. 3.14). There are some screws with hexagonal recesses in the head but they are uncommon in agriculture.

Most common sizes of key are from $1/8$ in to $1/2$ in or from 3 mm to 12 mm and a set to fit in this range will be adequate for the farm workshop. Bigger keys can be bought separately.

Measurements are taken across the flats of the hexagonal. Always use the correct size of

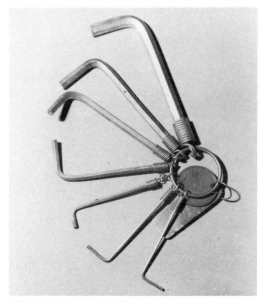

Fig. 3.13 A set of Allen keys is not expensive.

Fig. 3.14 Grub screws usually need Allen keys.

key as it is easy to damage the flats of a grub-screw particularly the small ones.

A set of Whitworth, AF or metric sizes is not expensive and they can be bought attached to a key-ring.

FILES

Many jobs in the farm workshop can be done faster and better with a file than with a grinding wheel, but to get the best work from your files, a few rules must be adhered to. Always use a new file on soft metal such as brass or aluminium, then on slightly harder metal like cast-iron before introducing it to steel. This allows the teeth to 'temper' and makes them less prone to chipping or breaking if used directly on hard metals. Do not throw them in a heap with spanners and other tools; keep them in a rack or file wallet. A file is made from hardened tool steel and the sharp cutting edges on its surface tend to be brittle. When filing narrow surfaces or when a smooth surface is essential, use the draw-filing method (see Fig. 3.15). This also helps to avoid clogging of the file teeth.

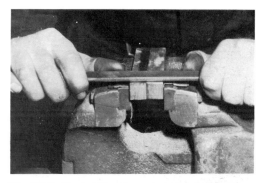

Fig. 3.15 Draw-filing to overcome clogging — an excellent way to file narrow surfaces. Pressure is applied to and fro.

When filing use slow full-length strokes and do not allow the file to skid over the metal. Short fast jerky strokes do not give the teeth time to cut and the work is likely to vibrate and cause screeching. If used in this manner a file will cut unevenly and soon become dull.

Fig. 3.16 (Top to bottom) 14 in flat bastard (30 teeth per inch) mainly for rough work; 12 in second cut (40) with teeth running diagonally for smoother work; 12 in bastard with teeth running diagonally in two directions; 10 in second cut for finer control; 6 in smoothing file (50) ideal for sharpening mower knife sections.

Files are classified according to teeth, size, shape, use and coarseness of cut — rough, coarse, bastard, second cut, smooth and dead smooth. These terms vary with the size of file. For example, a 20 cm bastard file is finer than a 25 cm. A file with one series of teeth is known as a single cut and with two rows, which cross at an angle, a double cut (Fig. 3.16).

Once a file has been used on steel it will be less efficient on soft metals and it is good practice to keep one side for soft and the other for hard metals.

It is as important to look after files properly as it is to know how to use them correctly.

A clogged file will not cut properly; it will probably scratch the surface instead. A wire brush should be used to clean files and rubbing on chalk is said to minimise clogging (Figs 3.17 and 3.18).

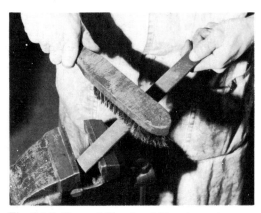

Fig. 3.17 Clogged files should be cleaned with a wire brush.

Fig. 3.18 Rubbing chalk into the file keeps clogging to a minimum.

To store files a simple wallet can be made by folding like a concertina, thick paper, cardboard or similar material, placing a file in each fold, bunching them together and tying with string.

Screw slotted angle-iron to a wall so that the slots form a rack for different sizes of file (Figs 3.19 and 3.20).

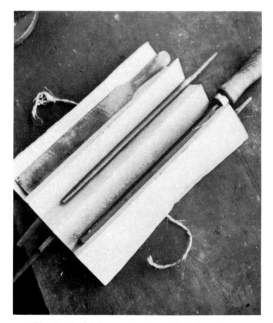

Fig. 3.19 Make a simple wallet for them.

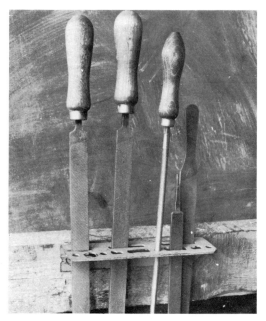

Fig. 3.20 A piece of 'Dexion' type angle iron makes a good rack.

PUNCHES

The farm workshop should be equipped with three or four kinds of punch for removing pins, marking out jobs and for alignment purposes (Fig. 3.21).

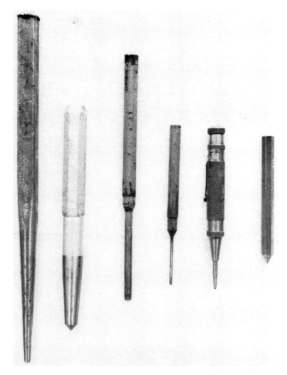

Fig. 3.21 A selection of punches. From left: taper, centre, parallel or pin punches (two), spring-loaded centre punch made out of a broken Allen key.

Parallel or pin punches are used to remove locating pins and rusted bolts. Choose the correct diameter parallel punch when knocking out split-pins; too small a punch may become stuck between the two halves of the pin.

Centre or dot punches are used for marking the centre of holes for drilling or for identifying the centre of a circle for compass points. Grind the point to an angle of about 60°.

Buy two centre punches, a small one for initial marking and a larger one to make a bigger mark once the first one has been checked. Always hold a centre punch vertically or it may slide when hit, leaving an inaccurate mark.

A spring-loaded centre punch works in the same way as a percussion hammer. Press the punch handle with the point on the metal and it will slide down the shaft, overcoming a spring pressure inside the handle to make a mark.

Taper punches are useful for lining up holes when fitting tinware. Push the punch through the hole next to the one in which you want to fit a bolt and it will pull the tin together, with the holes in line for the bolt. Screw this bolt in loosely and go on to the next. When all bolts are in they can be tightened.

Wad or hollow punches are circular hollow punches with a cutting edge ground on to the work end (Fig. 3.22). Use them for cutting holes in home-made gaskets, canvas screens or other soft material. Never cut a hole in material lying on a steel surface; use a hardwood block between material and bench, to prevent damage to the cutting edge of the punch.

Fig. 3.22 A set of wad punches.

Punches need little attention apart from keeping them correctly hardened and tempered and properly ground. There should be no mushrooming on the handle and the point should be kept clean and free from burrs.

Spreading on the head of the punch is dangerous; bits may fly off when it is hit with a hammer and damage eyes or face. If the burrs are really bad through lack of attention over a long period place the punch head downwards on an anvil and with a hammer knock off the worst of the unwanted material. Clean off on a grind-stone.

Grind the point a little at a time frequently dipping the punch in water to cool it and prevent loss of hardness.

If the punch gets too hot and needs re-hardening use an acetylene torch to heat the metal to cherry-red heat along about a third of its length. Plunge it into water until cold. Clean the blade with emery cloth to see the colours formed by the re-tempering process.

Slowly reheat the punch. As the temperature increases the colours will begin to move slowly along the metal towards the point. When the edge is straw brown colour, plunge the point in water again.

Finally, lightly grind the punch and test for

hardness with one or two light blows before hitting it hard.

This process can be used to harden all cutting tools.

HACKSAWS

Many workshop jobs of repairing or manufacturing call for the cutting of metals with a hacksaw. It is tempting to use whatever blade happens to be in the frame, but this is asking for broken blades.

A wide range of blades is sold, with teeth pitch ranging from 32 teeth per inch (tpi) for sheet metal to 14 tpi for cast-iron.

Blades are designed to cut on the forward stroke, but it is possible to obtain those which cut in both directions. Blades are usually 12 mm wide and about 0.6 mm thick.

Some are hardened and tempered to the same degree throughout, but the best type for the farm workshop has hardened teeth with the rest of the blade softer. Another type, which is about 1 in wide, has teeth on both edges which are hardened while the centre of the blade is soft.

Both the soft-backed blade and the soft-centred blade are ideal for cutting in awkward corners. They are less likely to break.

Make sure that the teeth are in line with the frame. Blades tend to twist when tightening the frame thumbscrew, making it difficult to cut straight.

Hacksaw blades are like files in that they will last longer if used initially for cutting soft metals such as brass, copper or aluminium. Cutting against a sharp corner is the quickest way to blunt a new blade. The initial strokes should be made away from the edge. The hacksaw should be held by both hands (see Fig. 3.23). The index finger of the hand on the handle should lie along side of the frame, to help control direction. When the work is held in a vice, cut as close to the vice as possible.

A common fault is cutting too fast. This causes the teeth to get hot and lose their hardness; they do not have time to bite into the metal and will slide over the surface.

A blade with 18 tpi is best for all-round use. There should be at least two teeth completely in work at any one part of the stroke. Teeth should never straddle the work.

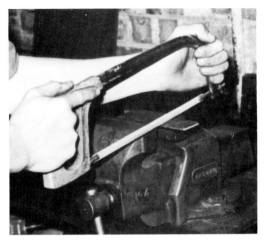

Fig. 3.23 Use both hands on the frame.

If thin sheet metal has to be cut by hacksaw and the only blade available is a coarse one, clamp several sheets together or clamp the sheet between wood.

Should a blade break halfway through a cut it is better to start again on the other side rather than force the new blade down the old groove.

BROKEN BLADES

Awkward jobs for hacksaws can mean more breakages than usual — and blades are not cheap. Mr. H. V. Hinton thought up this idea for using blades which are still good except for their lost inches. A piece of mild steel file cut to match the length of broken blade and shaped with the twin hooks one end is drilled to fit the hacksaw frame at the other. Hook on and tighten as usual (Fig. 3.24).

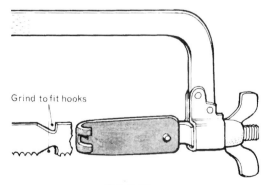

Grind to fit hooks

Fig. 3.24

COLD CHISELS

Cold chisels are made from high carbon or nickel alloy tool steel and are tempered for cutting cold metals. They must be kept sharp — and sharpening entails more than giving the edge a quick rub on the emery wheel. Hold the chisel in the right hand and keep it firmly against the grinder work rest with the index finger touching the underside of the rest. Press the cutting edge against the wheel with the first two fingers of the left hand and move it back and forth across the wheel with a wrist action. This produces a slightly curved edge by grinding more off the two corners, which helps prevent their breaking off and reduces the danger of drawing the temper through overheating (Fig. 3.25).

Fig. 3.25 Sharpening a chisel correctly.

To ensure that the cutting angle is correct, a simple gauge can be made from a small piece of sheet metal (Fig. 3.26).

Fig. 3.26 Gauge for chisel sharpening.

The harder the metal to be cut, the greater the cutting angle and vice-versa. A 40 degree angle is best for soft metals, 60—70 degrees for mild steel, and for hard metal, 75—80 degrees.

Chisels should be the right size for the job. Using a light one for heavy work may shatter it, and it will certainly vibrate and sting the hand. Figure 3.27 shows a selection of chisels suitable for the farm workshop.

When cutting or dressing metal on an anvil, always use the soft surface working face. Cutting right through a piece of metal onto the hardened surface will not only blunt the chisel, but damage the face of the anvil. The same goes for work held in a vice, keep the chisel away from the hardened jaws.

Never allow the chisel to become mushroom-headed because fragments may fly off when struck, also they can become dangerous to handle (see Fig. 3.28).

Chisels can be used for making very clean cuts in steel plate, provided the plate is held securely (see Fig. 3.29).

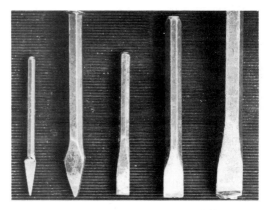

Fig. 3.27 (From left to right): Nos 1 and 2: cross-cut chisels for cutting rectangular grooves such as keyways. Nos 3, 4, and 5: flat chisels for chipping flat or convex surfaces and cutting in general. The damaged cutting edge of Number 5 is the result of continual overheating during sharpening.

Fig. 3.28 Do not let a chisel head become mushroomed.

Fig. 3.29 Cutting a steel plate.

HAMMERS

Hammers are probably the most commonly used tools in the workshop and also are the most abused. Often the wrong type is used for a job.

Hammers are classified according to shape of head and weight. Two or three different types will suffice for most jobs on the farm and in the workshop. Most common type in use in engineering is the ball pein, with one end of the head ball-shaped and the other end flat.

Peining is the act of striking a small area of metal a number of light blows with the object of stretching the metal slightly.

Other types of peining hammer are the cross pein and the straight pein, with a blunt-pointed head across the shaft of the handle and down the shaft respectively.

Other types of hammer include the claw, used for extracting nails, the bigger club or mason's hammer for heavy work and the sledge for a really big job (Fig. 3.30).

The 'soft' hammer or mallet avoids damaging a finished surface and yet can have some force applied, for example, when knocking a bearing on to a shaft.

Soft hammers are two headed, with one head of copper or lead and the other of hide. Both heads are easily replaced. For the big job use a sledge-hammer.

Hammer heads are usually made of medium carbon steel. The striking ends are hardened for long life and the middle portion, where the handle fits, is left soft. Never strike a blow with a middle portion which will easily lose its shape, causing the handle to work loose.

Hammer handles are made of metal or, more commonly, of wood, such as hickory or ash for toughness and strength combined with cheapness.

Make sure the handle is tight and free from cracks. When fitting a new one, always use two wedges — one wood, one metal — to hold the handle on the head. The wooden wedge should be driven in with the grain of the wood in the groove provided in the end of the handle. The metal wedge goes across the grain and helps

Fig. 3.30 From left: 2 lb ball pein, 1 lb cross pein, copper and rawhide soft hammer, mason's hammer, 1 lb claw hammer, ½ lb ball pein and a 14 lb sledge hammer.

hold the wooden one in position.

Use a 2 lb or 3 lb hammer for a big chiselling job such as hacking a rusted nut off a shaft. For a delicate job, such as tapping out a gasket, use a half-pounder.

Grip the hammer at the end of the shaft to exert the maximum force and use an easy rhythmic swing on to the surface to be stuck. Take an aiming tap between each main blow until you are sure of your aim.

RIVETS

Most components of modern farm machinery are fastened into position by bolts, spot welding or dowels, but there are still instances where the rivet is used to join parts.

Fig. 3.31 Three main types of rivet, from left: round head rivet, countersunk and pan head, which is the strongest type.

Three main types of rivet are used (Fig. 3.31). One is the countersunk rivet when a flush surface is needed. Because part of the main strength of a rivet is in the head which prevents components pulling apart, this type is not as strong as the other two — the pan head and the round head. The round head is rounded and the pan head is bulkier — the strongest of the three.

The strength of the finished job also depends on the tightness of the rivet in the joint. For this reason it is essential to use the correct size. If it is too small the rivet will not spread in the hole sufficiently and will allow movement. The strain on the rivets will be too great and they will

shear. The correct sized rivet should pass comfortably through the hole and when expanded, grip the metal of the parts that are being joined.

If the rivet holes have worn oval they must be welded up and redrilled or a loose joint will result.

Length of rivet is also crucial. If it is too short the rivet will not protrude far enough through its hole to allow a strong head to be formed. Aim to have about 1½ times the diameter of the rivet showing above the surface.

To make a neat job of forming the head use a rivet snap. This is a high carbon steel 'punch' with a semi-circular recess in one end. This is placed over the rivet and the other end is hit with a hammer. The rivet head takes the form of the recess. There is a range of snap sizes to form the right head on any common sized rivet (see Figs. 3.32 and 3.33).

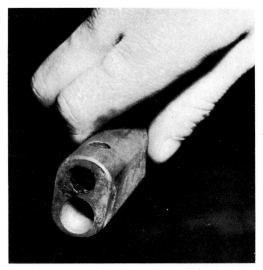

Fig. 3.32 This snap is used first to tighten the joint, then to form the rivet head.

There are two basic methods of riveting: hot riveting that entails heating the rivet up to a cherry red heat before placing it in the parts to be joined, and cold riveting. Usually hot riveting is used only on larger rivets as it is easier to form the head of the rivet and to get a better finished job. As it cools the metal contracts, pulling the surfaces together to give a tight joint. Cold riveting is used generally on smaller jobs such as a combine knife section.

Special rivets

Brake lining retention rivets are often hollow-

Fig. 3.33 The rivet snap. Hemispherical recess in the end of the snap is hammered over the rivet to form a strong head.

as fastening steel sheets to gates (Fig. 3.36). They are often better than welded joints because they do not cause the steel sheet to distort and there is no chance of burning off the zinc coating when galvanised sheets are being used.

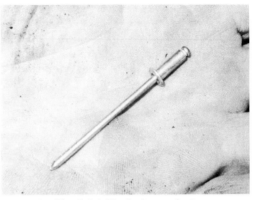

Fig. 3.34 Typical pop rivet.

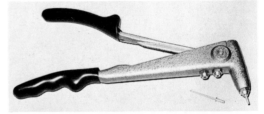

Fig. 3.35 Pop rivet gun.

Fig. 3.36 Galvanised steel sheet pop riveted to a door.

shanked and made of copper. When they are inserted it is usual practice not to attempt to form a second head but merely to flatten the end of the rivet after a punch has been used to expand the hollow end.

Do not be tempted to over-flatten the rivet as this will reduce the strength of the head. Rivets made of other materials may be available for this purpose, but it is preferable to use copper for brake linings. Should the linings become worn down to the rivet heads, there is less chance of scoring the drums.

Bifurcated rivets are used for soft materials such as plastics and leather.

Pop rivets were invented because the aircraft industry needed a rivet that could be formed in holes where access to the back was not available.

Each rivet is tubular and is usually made of an aluminium alloy. To expand the rivet a steel rod runs through the centre and is lightly secured for convenience. The rod has a small head which is pulled into the rivet and causes the open end of the rivet to bell out. Having expanded the rivet, the rod snaps off just behind the head leaving a neat flat head on the exposed side and a bell-mouthed head on the hidden side (Fig. 3.34).

A special pop rivet gun is required. (Fig. 3.35).

Pop rivets are available in standard sizes. The size of hole drilled should be similar to the rivet size.

Pop rivets are cheap and ideal for jobs such

FEELER GAUGES

Feeler gauges are used to measure critical clearances such as tappet, contact breaker or plug gaps, or to determine the number of shims required for bearings or cutter bars. You will probably need both an imperial and a metric set.

The gauges are made of hardened steel and there are generally 10 in a set, ranging from 1½ to 25 thousandths of an inch thickness; in imperial measure or in metric units feelers are made in hundredths of mm and range from 5 hundredths of 1 mm to 100. A set of sizes 5 to 60 would cover most likely applications (Fig. 3.37).

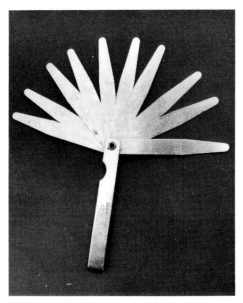

Fig. 3.37 Typical set of feeler gauges.

They are precision instruments and should be treated as such. When using them to measure a tappet clearance, for example, do not tighten the tappet hard down on the gauge. Adjust for a light dragging fit between the tappet and valve. A tight fit will scratch or dent the surface of the gauge, so making it inaccurate.

After use, clean the gauges, and store them in rust-inhibiting paper, or smear them with light oil, wrap in an oily rag and store in a dry place.

Rusty gauges can be startlingly inaccurate, especially when two or more are used together.

MICROMETER

When measuring critical dimensions such as the diameter of a shaft or the width of a key, a micrometer gives extreme accuracy.

An external micrometer consists of a frame with a spindle and a sleeve at one end and the anvil or fixed jaw at the other. The sleeve has an internal thread and calibrations on the out-side. The spindle is screwed into the thread and moved by turning the thimble on the barrel (Fig. 3.38).

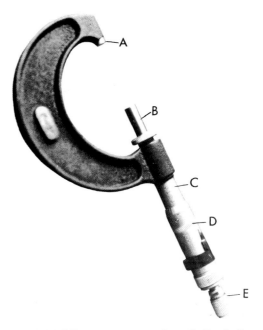

Fig. 3.38 Micrometer parts: A anvil, B spindle, C sleeve, D thimble, E ratchet.

Place the jaws of the micrometer over the object to be measured and, by turning the thimble, move the spindle until the jaws lightly grip the object. Read the measurement off the sleeve.

It is important that the jaws are not tightened too much, otherwise they will become strained and the extra pressure put on the fine thread inside the sleeve may damage it.

To ensure that the jaws are not over-tightened some micrometers have a ratchet attached to the thimble and this slips when a pre-set pressure is reached. Remember that using the ratchet to screw up the jaws gives greater accuracy.

For the farm workshop two micrometers should suffice — an ordinary 0—25 mm fixed anvil micrometer for small dimensions and an adjustable 0—100 mm for most of the bearings and shafts likely to be encountered on the farm (Fig. 3.39).

If your workshop is big enough to justify one, buy an internal micrometer as well. But a skilled mechanic will be able to use a pair of internal calipers and an ordinary micrometer to measure internal diameters.

Fig. 3.39 Measuring the diameter of a crankshaft. Remember to use the ratchet to screw up the jaws.

Metric micrometer

The screw has a pitch of ½ mm; therefore two revolutions of the thimble will move the spindle a distance of 1 mm.

Sleeve
The datum line is graduated with two sets of lines, the group above reading in millimetres, the group below in half-millimetres.

Thimble
The scale is marked in fifty equal divisions, in groups of five, each small division representing 1/50 of ½ mm which equals 1/100 mm (0.01 mm).

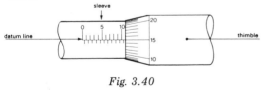

Fig. 3.40

To read the micrometer (Fig. 3.40)
(1) Read the number of whole millimetre divisions on the sleeve (major divisions).
(2) Observe whether or not there is a half-millimetre division visible (minor divisions).
(3) Read the thimble for hundredths (thimble divisions), i.e. the line on the thimble coinciding with the datum line.

Major divisions	10 x 1.00 mm	= 10.00 mm
Minor divisions	1 x 0.50 mm	= 0.50 mm
Thimble divisions	15 x 0.01 mm	= 0.15 mm
	Reading	= 10.65 mm

Imperial micrometer

The screw has 40 threads per inch so one complete revolution moves 1/40 in (0.025 in) and in 1/25 of a turn it will move 1/25 of 1/40 in (0.001 in).

Sleeve
This has the major divisions marked on it representing tenths of an inch, i.e. 0.100 in. Each major division is sub-divided into four minor divisions representing 0.025 in each.

Thimble
This is divided into twenty-five parts and as one full turn is equal to one minor division on the sleeve (0.025 in) each division on the thimble will be 0.001 in.

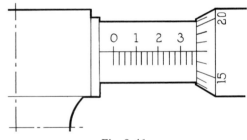

Fig. 3.41

To read the micrometer (Fig. 3.41)
(1) Read the number of tenths (major divisions).
(2) Add the number of minor divisions multiplied by 0.025 in.
(3) Add the number of thousandths on the thimble, i.e. the line on the thimble coinciding with the datum line.

Major divisions	3 x 0.100 in	= 0.300 in
Minor divisions	2 x 0.025 in	= 0.050 in
Thimble divisions	17 x 0.001 in	= 0.017 in
	Reading	= 0.367 in

Variable micrometers

There are two types of external micrometers. One has a fixed anvil and the other has a removable anvil. With the removable type one can get a set of anvils in varying lengths (0—10 cm or 0—15 cm) and by changing the jaw can measure a wide range of sizes with one micrometer (Fig. 3.42).

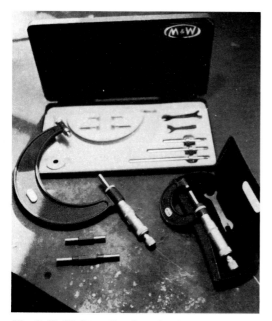

Fig. 3.42 Right: a fixed anvil micrometer.
The larger, variable type is on the left.
Interchangeable anvils are in the box, and the
master gauges are bottom left.

To change the jaws a set of spanners is
provided with the micrometer. Take out one
jaw and substitute another, making sure it has
seated correctly. Before use check the micrometer for accuracy with a master gauge or a
piece of known size, in case the new jaw has
not seated or dirt has stopped it screwing in
fully.

Check with a master piece of 5 cm diameter;
if the micrometer reads zero when the master
piece is measured then all is correct. If it reads
0.2 mm over size, on all readings you will have

to subtract 0.2 mm to give an accurate measurement and vice versa if the reading, when checked
on the master piece, is low.

Never swap parts of one micrometer with
parts of another; each micrometer is made and
calibrated separately so that it is only accurate
with its own parts.

Cleanliness is essential for good results.
Before a measurement is taken, clean off all dirt
and grease from the part to be measured or a
false reading may be obtained.

Do not ill-treat the micrometer by giving it
a hard knock, or its accuracy will soon be lost.
When it has been used, clean it with a dry rag
and store it in a dry place with a layer of rust-
preventing paper.

When the zero reading is not correct, the
error must be added to or subtracted from the
measured reading.

The error may be eliminated by rotating
the barrel until the datum line coincides with
the zero on the thimble when the anvils are
together. This is achieved with the aid of a
C-spanner (Fig. 3.43) supplied with the instrument. When necessary the anvils of a larger
micrometer can be adjusted individually.

Fig. 3.44 Measuring an internal diameter with
a telescopic gauge.

Fig. 3.43 Adjusting the barrel to correct zero reading.

TELESCOPIC GAUGES

Measuring an internal dimension such as the base of a bearing or cylinder can be overcome in several ways.

The ideal method is to use a telescopic gauge (Fig. 3.44). This has a spring-loaded plunger which is pushed into the gap to be measured, allowed to expand, then locked by rotating the handle. The gauge is then withdrawn and a micrometer used to measure the distance across the ends of the plungers (Fig. 3.45).

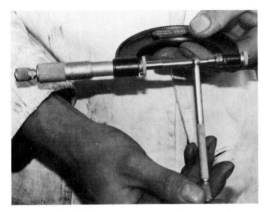

Fig. 3.45 Measuring across the ends of a telescopic gauge.

As telescopic gauges are expensive, a cheap pair of vernier calipers may be used instead.

The caliper can also be used as an accurate depth gauge. In many instances a good-quality vernier caliper will be accurate enough for the farm workshop. Usually such calipers can be read to the nearest 0.001 in (Fig. 3.46).

Fig. 3.46 Measuring an internal diameter with a vernier caliper.

There are occasions, such as when setting the valve stem clearances of an overhead cam engine, when shims 0.005 in larger or smaller than the original must be substituted.

One way of doing this is to obtain an adjustable wrench which has parallel jaws. Suppose a shim which is 0.005 in larger than the original must be selected from an assortment of shims of unknown thickness.

Clamp the original shim in the adjustable wrench together with a 0.005 in feeler gauge on top of it, then select another shim which will slide between the jaws of the spanner alongside the original, plus the feeler gauge.

Fig. 3.47 Spanner and feeler gauge to measure the shim thickness.

Figure 3.47 shows the thickness of a forage harvester knife shim being measured.

TORQUE WRENCHES

All bolts and screws on machinery need to be tight — but how tight? The torque wrench enables bolts to be tightened or 'stretched' correctly by measuring how hard they are to turn. Threaded components need to be in good condition because damaged threads will make a bolt abnormally difficult to turn. The wrench will record the effort needed to turn the bolt but will not apply the amount of stretch necessary to hold machinery components tightly together.

Torque is the turning effect or twist on the bolt produced by the length of the spanner used and the pull on the end of it, expressed in pounds-feet. A torque of 50 lb ft is achieved by exerting a pull of 50 lb on the end of a spanner 1 ft long, or 25 lb at the end of a 2 ft spanner. Strictly speaking, a pull of 1 lb should be written 1 lb force (1 lbf), and torque expressed in 1 lbf ft, the unit used here.

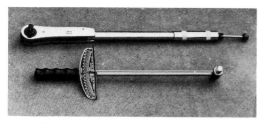

Fig. 3.48 Modern torque wrenches.

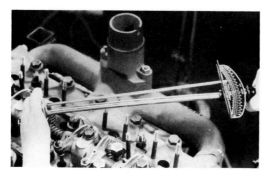

Fig. 3.49 Tightening a cylinder head.

Figure 3.48 illustrates two modern torque wrenches and Fig. 3.49 shows a typical torque wrench being used to tighten cylinder-head bolts. This type has a pointer which indicates the torque being applied. On some other types, a click is heard when a pre-set torque has been reached. Modern torque wrenches are usually calibrated in pounds force feet (lbf ft), kilogramme force metres (kgf m) and newton metres (Nm). Eventually, all British wrenches will be calibrated only in newton metres, the newton being the SI unit of force or pull. One pound pull is equal to about 4.5 newtons. British workshop manuals usually state the torque in lbf ft, plus one other unit. To calibrate an old torque wrench in modern units, use the following table:

To convert:
lbf in to kgf m, multiply by 0.012.
lbf ft to kgf m, multiply by 0.138.
lbf in to Nm, multiply by 0.113.
lbf ft to Nm, multiply by 1.136.

When using a torque wrench always make sure that the correct torque figure for the bolt being tightened is known. Do not guess.

The torque figure in the workshop manual will usually apply to clean, dry threads. Lubricate the threads if the book says so. Do not over-lubricate for blind holes in castings, or the excess oil may form a hydraulic lock which could crack the casting.

Tighten components such as cylinder heads in the correct sequence shown in the workshop manual. When the correct sequence is not available, work outwards from the centre of the cylinder head. Tighten in several stages. If the correct torque setting is 60 lbf ft, tighten to 35 lbf ft, then 50 lbf ft and finally to 60 lbf ft.

If a torque wrench is suspect, check it as follows: Suspend the wrench horizontally from a vice (Fig. 3.50).

Fig. 3.50 Testing a torque wrench.

Hang a weight of, say, 20 lb on the handle at a known distance (say 2 ft) from the drive end and check that the wrench actually records 40 lbf ft of torque, plus its own weight. Alternatively, the downward pull could be provided by a spring balance.

The torque wrench is a precision tool and should be looked after for it to remain accurate. Loosen the spring after use on the pre-set types, or it will become weak and the wrench will tend to over-read.

Few people can accurately guess when a bolt is correctly torqued. Without a torque wrench, check the torque as follows:

● Mark the spanner being used with a convenient multiple of a foot, i.e., 6 in (0.5), 12 in (1), 18 in (1.5).

● Obtain a suitable spring balance and attach this to the spanner, as shown in Fig. 3.51.

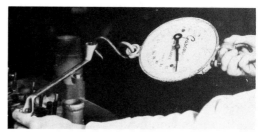

Fig. 3.51 Improvising without a torque wrench.

• Pull on the spring balance until the correct torque is obtained. The pull must always be at right angles to the spanner or the bolt will not be tightened enough.

Using this method, a 9 in-long spanner (0.75 ft) with a pull of 60 lbf will result in a torque of 60 x 0.75 = 45 lbf ft. When only a small spring balance is available, a piece of pipe slipped over the spanner will greatly reduce the reading required on the spring balance. In this case, with the spanner extended to 18 in (1.5 ft) the reading on the balance would be reduced to 30 lbf.

DRILLS

Of the many types of twist-drill for drilling metal, the most common are the parallel-shank jobber series, stub drills and the morse taper-shank series.

The parallel-shank jobber drills range from 9.2 mm to 16 mm in diameter.

Stub drills are shortened versions of the parallel-shank jobber drills, and have a maximum diameter of 25 mm (about 1 in). They are more suitable for most farm work and less likely to break.

Morse taper-shank drills are used for large holes (Fig. 3.52). They require special machines because the drill is held and driven by its taper. Special morse-taper sleeves tailor each size of the drill to the machine (Fig. 3.53).

Fig. 3.52 A typical morse taper-shank drill. It requires a special machine because the drill is held and driven by its taper.

Fig. 3.53 Morse taper drill and sleeve.

Drills vary not only in length and type of shank but in having quick or slow helixes to suit the material being drilled. The 'quickness' of the helix decides the rake of the drill. 'Rake' is similar to the pitch of a plough share.

Fast helix drills are usually used for soft metals, such as aluminium. Slow helix drills are used for materials such as brass, which require a slight rake, or when drilling thin metal to prevent the drill 'corkscrewing' as the point breaks through.

Should the drill corkscrew, the thin plate being drilled away may be torn from its clamp and rotate with the drill. This is dangerous and can break the drill. When no slow helix drill is available, the two cutting edges of a normal drill may be ground back in order to reduce rake.

When a large number of existing holes have to be enlarged, use either a standard drill with its cutting edges ground back or preferably a three- or four-flute drill. These drills have slow helixes and the extra flutes enable them to operate more smoothly than a standard two-flute drill.

Always keep drills sharp. As soon as any excess force is needed to make the drill cut, stop work and sharpen the drill. Use a cutting fluid, such as a soluble oil, when drilling steels.

Cast-iron and brass should be drilled dry. Keep the drill cutting. Never allow it to rub on the bottom of the hole or it will overheat.

Always use the drill at the correct speed. There is a tendency to use small drills too slowly and larger ones too fast. For instance, when drilling mild steel with a 3 mm diameter high-speed steel drill, the spindle speed can be 3000 r.p.m. But increase the drill size to 12 mm and the speed must be reduced to about 750 r.p.m. Plain-carbon steel drills should be used at no more than half these speeds.

Plain carbon steel drills are cheaper than high-speed steel drills and will last a long time when used at the correct speed. The smaller sizes are not so likely to snap as the steel is less brittle.

Before starting to drill a hole, centre-punch a mark in the required place. Should the drill not follow the centre-punch mark, stop before the hole is too deep and carefully punch it in the desired direction.

Blunt drills cause problems and breakages, so accurate sharpening is vital. There are two main points to watch.

The cutting lips must have sufficient clearance behind the cutting edges to give the drill bite. The correct amount is about 12° (see Fig. 3.54).

The cutting edges must be the same length and make exactly the same angle with the centre

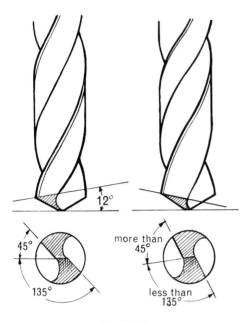

12°

45°

135°

more than
45°

less than
135°

Fig. 3.54

line of the drill. The angle for general work is 50 to 60°.

Cutting lips need enough clearance for the drill to bite into the metal, otherwise it will screech, vibrate and overheat. But if there is too much it will feed too fast, taking large bites at each revolution and increasing the possibility of breakage.

Clearance may be checked by standing the drill on a level surface, point down, alongside a steel rule and then turning it slowly. If the heel of the cutting edge is slightly higher than the front of the cutting edge and the angle is about 12°, clearance is correct (see Fig. 3.55).

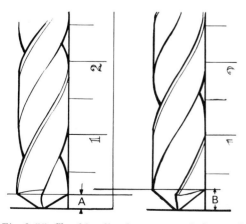

Fig. 3.55 Checking lip clearance angle by steel rule. B should be slightly greater than A after rotating the drill.

Drills need different cutting angles for different metals. For all-round mild steel drilling 55 to 60° is best; for cast-iron 50° and for brass 45°. Hard steels are difficult to drill but by grinding the cutting angle flatter to 75°, the job may be made easier.

If both cutting edges are not the same length the point of the drill is not central. It will bore an oversized hole and transfer most of the work on to one edge causing it to become blunt more quickly. If the lips are not ground to the correct angle the drill will be either too blunt or too pointed.

To check the angle, put two bright steel hexagon nuts together. The adjacent pair of flats give an angle of 120° thus making it possible to check both edges together (see Fig. 3.56).

Fig. 3.56 Checking the angles of the cutting edges.

Before offering the drill up to the grinder adjust the workrest to about the same level as the centre of the wheel. Then place the drill on the rest so that it lies horizontal (Fig. 3.57) and

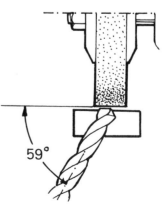

59°

Fig. 3.57 Correct position for grinding cutting angles.

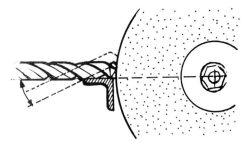

Fig. 3.58 Correct position for grinding clearance angles.

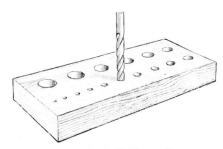

Fig. 3.59 Drill stand.

with the cutting lip level. It should also be at an angle of 55 to 60° with the face of the wheel (Fig. 3.58). A guide line cut into the rest may help.

The cutting edge should then be pushed against the wheel and slowly raised by pushing down on the shank using slow deliberate strokes. The drill must be frequently cooled in water to prevent overheating and 'drawing' the temper.

After sharpening, the web between the lips should be central and the cutting lips should be of equal length and the same angle. Viewed from the side, the cutting edges of the drill ideally should have equal and adequate clearances.

The thickness of the web increases towards the shank of a drill. As the drill becomes shorter, the length of the chisel edge between the lips becomes too long and excessive pressure is required to make the drill penetrate — usually when a drill is worn to about half to two-thirds of its original length.

To overcome this, the point must be thinned by carefully grinding away some of the web preferably on a specially shaped grinding wheel.

Never use fingers to remove swarf from a drill — it is likely to be razor sharp.

Drilling machines should be fitted with guards to prevent swarf from causing injury.

Often small drills are bought and used only a few times before they are mislaid or lost because there is no proper place to keep them. Store your drills tidily and undamaged in this cheap, simple-to-make drill stand. An off-cut of a 10 cm x 2 cm plank and half an hour's work is all that is needed.

Mark a guide line 2.5 cm from both edges of the plank and drill a hole for each size of drill, working down from the biggest to the smallest. Cut to length, clean up any splinters and each drill will stand in its own hole for easy selection (see Fig. 3.59).

REAMERS

A reamer will produce a hole with a fine finish that is accurate to a thousandth of an inch.

The reamer is also used when the internal diameter of a bush has to be expanded so that a shaft fits with little clearance. Bushes leave little choice about the amount of metal left for the reamer to remove. But when finishing a drilled hole, the less surplus metal left, the better, provided that the drill has produced a reasonable surface.

There are many kinds of reamer. Those most likely to be used in the farm workshop are:

Fixed reamers
This tool will probably be bought for a particular job, such as reaming the bushes of tractor king pins. It is difficult to ensure that a reamer follows perfectly the axis of a drilled hole and the reamer should have a built-in guide known as a pilot.

Stepped reamers
King pin assemblies often have two bushes of different sizes which must be reamed in perfect alignment. An easy solution is to use a stepped reamer, which will ream both holes at once (Fig. 3.60).

Fig. 3.60 A stepped reamer will ream two holes at once.

Fig. 3.61 Typical adjustable reamer.

Adjustable reamers

An adjustable reamer with its conical pilot is capable of reaming any size hole within a small range. The tool shown in Fig. 3.61 will cover the range 1 15/64 in to 1 9/16 in. The conical pilot enables the tool to be guided accurately when two holes of different sizes have to be reamed in line.

This expanding reamer has a number of hardened, tapered slots. At each end of the blade is an adjusting nut. The effective diameter is changed by slackening one of the adjusting nuts and pushing the blades along their slots with the other.

Pushing the blades towards the deeper end of the slots decreases the diameter of the tool and vice versa. Some of these reamers are made with detachable pilot sections (Fig. 3.62).

Fig. 3.62 Adjustable reamer with removable pilot section.

Care of reamers

Reamers are expensive and proper care pays. After use they should be cleaned, lightly oiled and stored without touching any hard objects which could damage their cutting edges. Preferably wrap them in cardboard or store in a wooden box with compartments.

Use of reamers

A tap holder is probably the best way of holding a reamer for a positive grip and the holder also makes the correct alignment that much easier (Fig. 3.63).

Fig. 3.63 Adjustable reamer in use with conical pilots being turned by a tap holder.

Always turn a reamer in the correct direction. Most should be turned clockwise viewed from the shank end.

When a reamer is turned backwards the cutting edges are dulled. To withdraw it from a hole, continue to rotate it in a clockwise direction.

Never attempt to make drastic changes to the direction of a hole by pushing the reamer sideways. Excessive force will produce a poor finish and damage the reamer. The golden rule is to push lightly and keep turning in the correct direction.

Swarf produced by a reamer is extremely sharp, so take care.

GRINDSTONES

Maintenance on most farms involves sharpening forage or beet harvester blades and small tools, and a power grindstone will do the job quickly and efficiently.

The three main types are either bench-mounted, pedestal, or hand held (Fig. 3.64). For minor repairs and maintenance a bench-mounted grindstone will be adequate and cheaper.

Fig. 3.64 At the other end of the scale. This mini hand unit is ideal for getting into awkward corners.

Where major repairs and a lot of welding and fabrication is carried out the pedestal type is best as the extra power is useful for removing metal in welding preparation (Fig. 3.65).

Look for a machine that is well guarded and has an adjustable fence stop to prevent fingers and metal being dragged into the wheel. Some machines have a water reservoir fitted for cooling the object being ground. This is useful for

Fig. 3.65 Large industrial grinders are best for rough work. They may be picked up cheaply from engineering companies who are in the process of retooling.

cooling small tools but a bucket of water beside the machine is needed for cooling big jobs.

Spare grinding wheels are useful and will extend the range of the grindstone. If you can afford it buy several wheels of different coarseness to suit the different jobs you expect to do.

Nearly every farm workshop has a power grinder. These machines are potentially lethal and thousands of industrial accidents involving grinders occur each year.

The laws controlling the use of grinders for agricultural purposes are not clear, but where a farm has a full-time mechanic, or a workshop which could be termed a factory, the Factories Acts can apply and the 1970 Abrasive Wheels Regulations come into force.

These regulations state that anyone who misuses an abrasive wheel breaks the law. No one can fit an abrasive wheel to a machine unless trained to do so — which involves attending a one-day course at a technical college or similar institution.

In the eyes of the law, an abrasive wheel means any power-driven device consisting of abrasive material intended for grinding or cutting. This includes portable hand-held grinders fitted with grinding discs rather than wheels.

At present, farmers are not strictly bound to adhere to the Abrasive Wheels Regulations, but the Health and Safety at Work Act states safe methods of work must be adopted and every reasonable safety precaution taken — this includes self-employed persons. The 1970 regulations are a reasonable guideline to the use of grinders on the farm.

The 1970 Abrasive Wheels Regulations include the following points

The machine should carry a notice stating the maximum spindle speed.

The maximum spindle speed must be adhered to at all times.

Only trained personnel must be permitted to fit the replacement wheel.

Adequate guards must be fitted and maintained in good condition.

The correct type and shape of wheel must be used for the job in hand.

A means of cutting off the power to the machine must be provided in such a position that a person operating the machine can switch the power off.

Work rests must be correctly adjusted and be secure.

Cautionary notices regarding the regulations must be posted near the machine.

The floor around the machine should be uncluttered and prevented from becoming slippery.

No one must wilfully misuse any of the protection devices or guards.

The mounting flanges should be secure and ALL necessary paper washers must be correctly fitted.

It is the duty of every employed person to make use of guards etc., and to report any defect found to the appropriate person.

A grinding wheel is made up of abrasive grits and the bond. Each abrasive grit that does the grinding removes a small piece of the metal being ground. Aluminium oxide is commonly used for grinding high tensile steels; silicon carbide can be used for non-ferrous metals.

The softer the bond which holds the grits together the softer the wheel. A soft wheel is used for grinding hard materials because the soft bond allows worn grits to break away easily and expose fresh sharp ones. The ideal combination of bond and grit should be such that the grit is allowed to break away just as it becomes blunt. When a stone wheel is used for the wrong material, the used grits may not break away and the wheel becomes blunt — known as glazing. Alternatively, chips of the metal being removed may adhere to the surface of the wheel, causing loading. A glazed or loaded wheel will require dressing.

Wheels are often marked with a number of

Table 3.1

A	46	K	5	V
Abrasive	*Grit size*	*Grade*	*Density of wheel*	*Bond type*
A = Aluminium oxide	8—24 Coarse	Hardness of bond:	1—8 dense	V = vitrified
C = Silicon carbide	30—80 Medium	ABC→XYZ	9—12 open	This is most common
	80—180 Fine	←soft, hard→		

confusing symbols. In the absence of advice from the manufacturers, table 3.1 is a guide to the main symbols of British Standard marking system.

Selection of the grade of wheel depends on the type of work. All wheels have a grade number which represents the size of particles used to make it. For example, a grain of 36 means that the particles will pass through a screen with 36 meshes to the inch.

For high-speed work where a polished surface is not required, a 30 grain wheel is ideal, but for sharpening drills or chisels a medium fine 80 grain is best. It is a simple job to change wheels, and having two or three different grades is a good idea. This will overcome the problem of ruining a fine tool-sharpening wheel by having to use it for the rougher work.

Figure 3.66 shows a correctly fitted grinding wheel. Observe the following precautions:

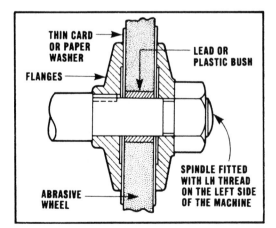

Fig. 3.66 Correctly fitted abrasive wheel.

(1) Examine the new wheel for obvious defects. Do not use a wheel that is chipped or cracked (Fig. 3.67).
(2) Test the new wheel for cracks by suspending it by a length of string and tapping it *lightly*

Fig. 3.67 A wheel chipped like this must be replaced immediately. It is out of balance and may disintegrate when it is used.

Fig. 3.68 Testing a wheel for cracks.

with a spanner (Fig. 3.68). A clear ring indicates a sound wheel. A wheel with a dull sound must be checked by an expert for flaws.
(3) Check that the wheel is suitable for the machine. It should fit easily, but not too loosely on the spindle. Too tight and the heat of grinding could cause the spindle to expand and

possibly crack the wheel.

(4) The spindle of the machine should run true on sound bearings and at the speed appropriate for the wheel being used. Maximum wheel speed should be marked on it.

(5) Nearly all farm grinders have wheels mounted between flanges that are not normally less than one-third the diameter of the wheel. They must be fitted with the recessed sides towards the wheel to ensure that the wheel is not clamped too tightly near its centre hole. The flanges must be the same size or there will be a tendency to warp the wheel.

(6) The paper washers supplied by manufacturers are important. Check that they are intact and slightly larger than the flanges.

(7) Do not overtighten the clamping nut — tighten it enough to prevent slippage. The nut on the left-hand side of the machine will probably have a left-hand thread.

(8) Adjust the work rest as near as possible to the wheel and rotate the wheel by hand to check that it is clear. The gap between the wheel and the rest should never exceed 3 mm (Fig. 3.69).

Fig. 3.69 Correctly adjusted work rest.

(9) Guards must be in position and properly adjusted.

(10) Make sure no other person is in the workshop, switch on and leave the grinder running for about quarter of an hour with no load. Should the wheel have a flaw it will usually disintegrate during the first few minutes. An incorrectly mounted grinding wheel could 'explode' during use.

For efficient work, the wheel must run true and have a sharp, clean face. To true a wheel, a diamond tool should be used. This is moved

slowly across the rotating wheel until all the high spots have been taken off. A piece of agate stone can be used but is less efficient.

Wheels which are running true, but have become dulled or loaded with the materials being ground, can be dressed with either an abrasive stick or a Huntington or similar type dresser. Figure 3.70 shows a Huntington dresser in use. This consists of a series of hardened steel washers moved across the face of the rotating wheel. The work rest has been moved away from the wheel to act as a guide for the dresser.

Fig. 3.70 A Huntington dresser.

Fig. 3.71 Misuse of a grinding wheel.

Should a wheel still vibrate after having been trued, it is probably out of balance and expert advice should be obtained before attempting to rebalance it.

Whenever possible, use the whole face of the wheel to keep it true. Do not use the side of the wheel. Figure 3.71 shows a wheel being dangerously misused. Firstly the wheel is being undercut and weakened; secondly, it is being stressed in the wrong place and could burst. Also, the flanges may be damaged and get out of balance.

Always keep the work rest properly adjusted. When the work rest reaches the end of its adjustment the wheel should be changed. Do not be tempted to elongate the slots in the work rest to extend the adjustment and do not use the machine with securing bolts missing from the work rest.

Wear goggles at all times, even when there is an eye shield on the machine. Goggles should be BS 2092 Grade I to give protection should a large particle of metal be thrown at the operator. Grade I goggles should withstand a ¼ inch diameter steel ball travelling at 300 mph.

Avoid wearing loose clothing which may get caught up.

Keep the floor area around the grinder uncluttered and free from oil and grease.

Never store fuel, combustible materials or batteries near a grinder.

Display a prominent notice — 'wear goggles' — next to the grinder.

Portable bench grinder

Grinders take up valuable bench space and often get in the way.

Mr John Fountain, machinery lecturer at Aylesbury College of Further Education, came up with this idea to overcome the problem (Figs. 3.72 and 3.73).

Fig. 3.72 The grinder can be used normally . . .

Instead of bolting the grinder to the bench top he mounted it on a base plate of 3 mm mild steel with a length of 50 x 50 x 6 mm angle-iron welded along the front edge.

The angle-iron holds the grinder in position and prevents it from being pushed on to the bench. The whole unit can be removed from the bench (Fig. 3.73) and taken to another part of the farm.

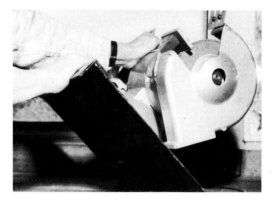

Fig. 3.73 . . . but is easily moved about.

CUTTING ROUND HOLES

A neatly-cut round hole looks more professional and does not weaken the metal as much as a ragged hole.

The best way to produce neat holes depends on the thickness of the metal, and the size and number of holes required. The tools and methods employed are:

Drill and file — Small holes can be produced by expanding a drilled hole using a half-round file.

Small items of thin metal sheet should be secured for drilling because as the drill breaks through there is a tendency for the workpiece to catch on the end of the drill and rotate with it.

Where possible, clamp the metal to a piece of wood before drilling. This will also produce a much neater hole. Do not try to drill a large hole in one operation. Always drill a pilot hole first.

Tank cutters — These will cut a neat hole, but each size of hole requires a separate cutter. One of the drawbacks of this method is that these cutters cannot be used easily in confined places.

Chain drilling

Mark the circle required then carefully drill a series of small holes around the *inside* of the circle, close together so that none touches the boundary of the circle. The waste material in the middle can now be removed by careful chiselling or, alternatively, by hack-sawing through a larger hole. The smaller the drill used for the chain drilling, the less work will be involved in finishing the job. If the holes run into each other, care must be taken to avoid breaking the drill.

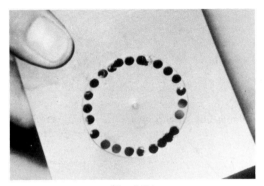

Fig. 3.74

Fig. 3.76

Nibbling tools

A monodex sheet metal cutter can be used to cut holes in sheet steel of varying thickness up to about 14 SWG. The blade in the end of the tool cuts a small slot by curling the waste material up in front of it. First a small hole must be drilled so that the blade can be inserted through the sheet.

Fig. 3.75

Fig. 3.77

Adjustable hole cutters

This hole cutter is operated by adjusting the cutter to the required hole size, drilling a small hole to position the centre of the tool and rotating it using a brace which allows a reasonable degree of force to be exerted on to the cutting tool. This tool could be modified to work with a power drill but rotational speed must be slow or the cutting tool may snap off.

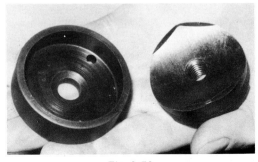

Fig. 3.78

Fig. 3.79

Fig. 3.80

Q-Max cutter

Where many holes of the same size have to be cut, especially in confined spaces in sheet steel, one of the best tools is the Q-Max cutter. First a hole is drilled large enough to take the centre bolt of the tool. Then the two halves of the tool are placed one each side of the sheet to be cut and the centre bolt inserted to locate the two cutting parts over the intended hole. To cut the hole, an Allen key is used to tighten the centre bolt so that the two cutting parts are drawn together. The part of the tool which the centre bolt screws into must be held station-

ary, preferably in a vice. The disadvantage of these tools is that again each tool will only cut one size of hole.

Oxyacetylene cutter

Neat holes can be cut with the cutting torch, provided the sheet of steel is thick enough to prevent distortion by the heat. A circle cutting attachment is essential. Several versions are available but all clip on to the cutting torch at varying distances from the nozzle. Normally, the centre of the proposed hole is marked with a punch hole in which sits the needle of the tool.

Set the torch pressure at about 0.3 bar acetylene and 2.0 bar oxygen and carry out a practice run on a similar piece of scrap material. Always position the sheet to be cut so that the hole can be made without having to stop several times. Rehearse the cut before lighting the torch to ensure that there is no danger of cutting through the gas pipes or that the pipes will not cause the nozzle to wander off course.

To make a clean start, either drill a hole at

Fig. 3.81

the circumference of the circle or, alternatively, blow through the sheet a short distance in from the circumference and then make a radial cut out to the circumference. Unless such a method of starting the hole is used, the hole may be too large at the initial blow through.

TAPS AND DIES

A tap is used to cut an internal thread and a die for the external thread. Both tools are made of hard metal and liable to break when used carelessly. A tap wrench and die stock are required in order to hold their respective components.

Taps are purchased in sets of two or three. The first tap, or taper tap, starts the threading operation. The first eight or nine threads are tapered to help the tap start in the hole. The second tap is tapered for only three or four threads, and is used to finish the thread in holes with open bottoms. In a blind hole, the third tap is used to finish. This 'plug' tap is tapered for only two threads at the most and therefore threads to nearly fully depth (Fig. 3.82). Blind holes should be drilled slightly deeper to allow the tap to clear itself.

Fig. 3.82 A taper tap on the right and a plug tap on the left showing the first few threads on the taper tap gradually leading to a full thread.

The size of hole required for a specified thread may be determined in several ways:

Table 3.2

Size in mm	Pitch in mm	Tapping size mm	Clearance size mm
6	1	5	6.3
8	1.25	6.7	8.2
10	1.5	8.4	10.2
12	1.75	10.3	12.3

(1) Refer to thread tables (table 3.2) and select the appropriate size of drill. Some thread tables list several tapping drills of different sizes to provide a range of thread engagement upon which is determined how close the male and female threads come to touching each other at their crests and roots.

(2) Try several sizes of drill in the new nut. The drill which will just pass through is the one to use as the tapping drill.

(3) Hold the tap against the light and place different sizes of drill behind it. The drill which covers all the core of the tap, but leaves the threads showing, is the one to use.

Unless the thread has to be gas or liquid tight, use the largest drill that will produce a decent thread. The larger hole will help the tap to adjust itself and reduce the risk of damage. Most drills tend to produce a slightly over-sized hole, unless they have been carefully sharpened.

Tapping a thread

Having first drilled the correct size of hole, fix the tap securely in the tap wrench, (see Fig. 3.83), then be sure to enter the taper tap at right angles to the work. Use a small set square to align it, (see Fig. 3.84). Once the tap has started, turn it clockwise for one turn then anti-clockwise for about half a turn to clear the threads. When tapping blind holes, withdraw the tap periodically and remove the swarf. A proprietary cutting compound will help when threading steel.

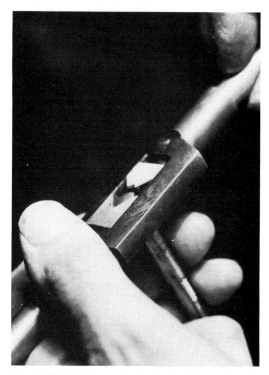

Fig. 3.83 Fix the tap securely in the tap wrench. If it is not held squarely, it will twist round with the action of tapping, and damage the soft jaws of the wrench.

Fig. 3.84 After the tap has begun to bite into the metal, check that it is entering the hole squarely. A lopsided thread is very difficult to correct after the first few turns.

Cast-iron is best left dry. Aluminium alloys, brass and copper may be lubricated with paraffin.

Should the first tap become tight, remove it and clear the hole with the second or plug tap, then re-insert the taper tap and continue. In soft materials, the tapping hole should be counterbored with a clearance drill to a depth of one thread, to prevent the material around the top of the hole swelling up because of the pressure of the tap.

Fig. 3.85 The adjusting screws to hold the die tight in the stock and to allow small alterations to be made in the size of the thread.

Soft or brittle materials should always be tapped with a coarse thread. A fine thread in these materials will be less strong.

Most dies to be found in the farm workshop will be of the split type (Fig. 3.85) which may be adjusted to vary the depth of the thread being cut. To cut an external thread, set the die to cut a shallow thread by expanding the die in its stock by the adjusting screw. Should the resulting thread be too large in diameter to enter the matching internal thread, close the die and run it over the thread for a second time until the external thread accurately fits a given size internal thread.

Therefore, when 'male' and 'female' thread are being made in the workshop, make the 'female' thread first because its size cannot be adjusted after it has been produced.

Cutting the external thread is similar to that for an internal thread. The die, like the tap, must be turned anti-clockwise regularly in order to clear the threads and the diestock must be held at right angles to the bar being threaded to avoid a 'drunken' thread, which will cause the

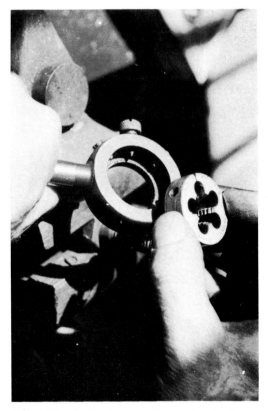

Fig. 3.86 The die fits into the stock with its identification marks farthest from the shoulder in the stock.

nut to wobble as it is wound along the screw. Chamfer the end of a bar slightly to help the die start squarely. Where a lathe or large drilling machine is available, use it to start the thread.

Hold the bar to be threaded in the chuck of the drill or lathe and rotate the chuck by hand while holding the die against the tailstock of the lathe or the table of the drilling machine.

A die must be used the correct way round. The leading end of the die usually carries the identification marks (Fig. 3.86). Used this way, the cutting teeth are less likely to bind because the trailing edges are farther from the bar being threaded than the leading edges. Dies used the wrong way round are likely to be difficult to turn and may easily be broken.

STUBBORN STUDS

Studs and bolts are often broken off in their threaded holes by using too much force on the spanner when tightening up. Removing the broken portion from the hole can present problems.

Stud extractors will remove studs protruding above the surface, such as cylinder head studs, but they are expensive and some damage the threads. A slower but cheaper method is to use two nuts locked on the thread one against the other. Use a spanner on the bottom nut and turn the stud out.

Several other ways of removal can be tried. Use a pair of grips and try to twist out the stud.

Failing that, weld a nut to the top of the broken portion.

If you have no nuts handy, weld a piece of bar across the stud and use it as a handle to turn the stud and remove it from the hole.

When the stud breaks below the surface try to undo it by turning it with a punch until it is above the surface.

If it is too far below the surface for this method drill a hole through the centre of the stud and use a sunken stud extractor, which is a steel tapered bolt with a coarse left-hand thread (see Fig. 3.87). It is held in a tap wrench and threaded into the hole through the stud (Fig. 3.88). Because of the taper the extractor tightens up in the hole and the left-hand thread will undo a right-hand thread stud or bolt.

If these methods fail, drill the stud right out and rethread the hole (Fig. 3.89). Find the exact centre of the stud and use a drill close to the tapping size of the stud thread. It is usually possible to refurbish the thread using a tap.

If all else fails drill out the stud with a drill which is the tapping size for a larger stud. Where it is essential that the stud size remains the same consider some form of stepped stud which will screw into the larger hole.

Whenever there is a choice use a coarse thread in weak material such as cast-iron. For this reason a steel cylinder stud is usually coarse-threaded at its lower end where it has to screw into a weak cast-iron block, but it is fine-threaded at its upper end as a steel nut is used to retain the cylinder head.

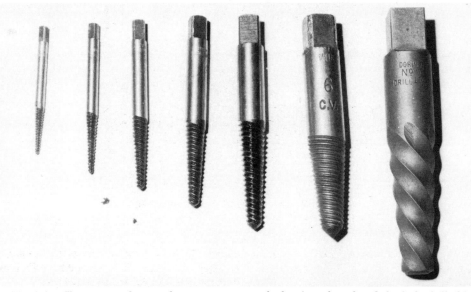

Fig. 3.87 Choose a sunken stud extractor to match the size of stud and the hole drilled in it.

Fig. 3.88 Insert the appropriate extractor and twist anti-clockwise.

Fig. 3.89 Drill right through the stud, preferably using a vertical drilling machine. Drill a small hole first.

A fine-threaded nut is less likely to work loose than a coarse thread and can be tightened with less effort.

There are several releasing fluids on the market. These in time penetrate the threads of a rusted nut. The usual difficulty with these fluids is that they do not act quickly.

But a squirt of releasing fluid followed by

shock treatment can often loosen a stubborn nut.

Any nut which refuses to move after shock treatment should be dealt with by using gas welding equipment or a blow lamp. Heat applied directly to a nut will expand and often crack the rust seal. Heat followed by quenching will often free the nut.

Some components can be damaged by heat — aluminium can melt and cast-iron crack. There is also a fire risk, especially when a nut being heated has already been treated with releasing fluid. Keep a fire extinguisher to hand. When treating wheel nuts in this way take care not to overheat the tyre.

The next form of attack against a rusted nut is a cold chisel. First try to unscrew the nut by tapping round it with the chisel. Aim to rotate the nut. But remember that the chisel will also stretch the nut as it cuts into it.

Should the nut still be stuck, use the chisel to split it across one of its flats and not down its length which may damage the threads. Most nuts do not need to be broken completely and can be removed after a small groove has been cut in one flat (Fig. 3.90).

Fig. 3.90 Cutting the nut with a chisel.

Fig. 3.91 Hacksaw a cut in the nut.

Where a nut is accessible remove it by cutting into two with a hacksaw. Again there is no need to break the nut completely. It will usually unscrew when sawn halfway. Be careful not to damage the threads (Fig. 3.91).

Should other methods fail, cut off the nut

Fig. 3.92 Gas cut the nut.

with a gas torch. This can be done without damaging threads (Fig. 3.92).

THREAD REPAIR

External threads can be repaired with an ordinary die, but many other devices are available. They include:

Fig. 3.93 Die nut.

Fig. 3.94 Thread file.

(a) *Die nuts* which are similar in appearance to cutting dies, but are not split. This can be turned with an ordinary spanner and no die stock is required, (see Fig. 3.93). It is intended for repairing threads and should never be used for cutting a thread.

(b) *Thread files*. These special files can be used to repair eight different pitch threads. Used

Fig. 3.95 Re-threading tool.

skilfully, the tool can clean threads which have been quite badly damaged (see Fig. 3.94).

(c) *Re-threading tools.* These are expensive but cover a wide range of thread, diameter and pitch. They can be used when a large diameter die is not available (see Fig. 3.95).

Internal threads can be repaired by using a tap, but sometimes they may have to be repaired by specialists who drill out the damaged threads and then tap the hole larger to take a special insert known as a helicoil. These inserts are effective and can be fitted quickly at a cost normally much less than replacing the component (Fig. 3.96).

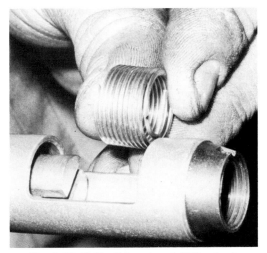

Fig. 3.96 Helicoil insert and fitting tool.

Helicoil inserts are often stronger than the original threads and can be used in most engineering jobs including those in which cylinder head studs or bolts are fitted into cylinder blocks made of aluminium.

THREADING PIPES

The weakest parts of galvanised steel water pipes are the joints. They are usually composed of an external thread on the pipe screwed into the internal thread of the fitting.

British steel pipes carry a thread known as the British Standard Pipe Thread (BSP), sometimes called a 'gas thread' or 'national gas thread', which has an angle of $55°$ and is usually tapered.

Taper-threaded pipes from the United States will probably carry the American Standard Taper Pipe Thread, or National Pipe Thread (NPT) which is the old name. NPT threads have an angle of $60°$ and very few of them are interchangeable with their BSP counterparts.

Pipe threads are based on the bore diameter. A pipe with nominal ½ in bore size will have a ½ in BSP thread. The actual outside diameter of the ½ in BSP thread will be 0.825 in. But the exact bore of the pipe may not be ½ in. This will depend on thickness of the wall. Nominal size means standard size, but in practice there can be small deviations.

To make a threaded pipe joint, a taper pipe thread is cut on the outside of the pipe to match the female thread in the fitting which can also be tapered. On the farm, the external thread is usually cut with a special stock-and-die set (Fig. 3.97). The stock carries four separate thread cutters, often called chasers, which can be expanded or contracted according to size of pipe. They are all moved the same amount by a cam plate (Fig. 3.98).

Fig. 3.97 Typical pipe threading stock and die set.

Another set of chasers must be fitted to cut different numbers of threads to the inch. Chasers can often be changed without removing the

Fig. 3.98 Cam plate removed for inspection.

Fig. 3.99 Initial setting mark.

cam plate.

The four chasers are used as a set and they must be fitted so that the cutting edges form a helix. They are numbered and must be fitted in the matching number on the holder.

Table 3.3 Common BSP threads

Size	Outside diameter	Threads an in	Tapping drill
⅛ in	0.383	28	0.3445
¼ in	0.518	19	0.4646
⅜ in	0.656	19	0.6004
½ in	0.825	14	0.7500
⅝ in	0.902	14	0.8281
¾ in	1.041	14	0.9646
1 in	1.309	11	1.2106
1¼ in	1.650	11	1.5551
1½ in	1.882	11	1.7812

Fig. 3.100 Information stamped on a chaser.

Table 3.3 shows the BSP thread sizes. The same set of chasers can be used to cut threads on ½ in, ⅝ in and ¾ in pipes because all have 14 threads to the inch. Similarly ¼ in and ⅜ in pipe threads have 19 to the inch and can be cut with the same chasers. Do not mix sets of thread chasers.

Adjustable pipe-die sets have reference marks on the cam plate which, when lined up with the appropriate mark on the holder, provide an initial setting for a standard size of thread (Fig. 3.99). The exact setting for the cam plate is then determined by cutting a test thread with the chasers in the closed position.

To thread a pipe, first determine the nominal size of the pipe and examine the chasers to decide which set to use. Usually this information is stamped on the chasers (Fig. 3.100). If not, consult the BSP Table.

Set the cam plate to the corresponding reference mark, close the chasers and place the die holder on the pipe, guide first, so that the pipe enters the tapered end of the chasers.

Adjust the guide to suit the pipe and push the holder towards the pipe. Rotate it clockwise until the chasers take hold. There is then no need to apply pressure because the tool will feed itself. Use a cutting compound such as lard oil.

Unlike cutting machine threads, there is no need to reverse the die. Cutting can be continuous because there is ample room between the chasers for the swarf to escape. As the number of threads cut on the pipe is the same as on the chasers, this reduces the need to clear the chasers by reversing them.

Continue rotating the holder until the pipe

just protrudes through the chasers, then open the chasers slowly as they are rotated another half turn (Fig. 3.101). This avoids a burr where they stop.

Fig. 3.101 Open the chasers while rotating them.

Until the dies have been correctly set for a particular pipe size, they may cut a thread too deep or too shallow.

Test the thread by screwing on a fitting. You should be able to turn it three times by hand before it becomes tight. Once the fitting is hand-

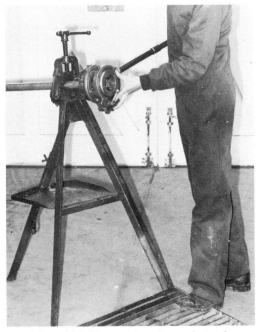

Fig. 3.102 Strong pipe vice and stand.

tight, two or three turns with a wrench should produce a watertight joint.

A pipe vice is needed to hold the pipe being threaded. The type shown in Figure 3.102 is portable and stable because the operator can stand on the base.

Fig. 3.103 Pipe cutting tool in action.

A useful tool for cutting pipes (Fig. 3.103) clamps on to the pipe and is rotated while the handle is tightened. This cuts through a thick pipe with much less effort than a hacksaw and always cuts the pipe square. This tool throws up a small shoulder on the outside and a burr on the inside of the pipe which must be removed before threading.

A damaged thread in a fitting or coupler can be repaired with a pipe-thread tap. This cuts an internal thread so that a tapered pipe thread can be screwed in.

Pipe-thread taps have the first few threads ground away to make starting the tap easier. The depth to which the tap should be screwed can be estimated by screwing the mating pipe into the fitting. A few threads should still be showing when the pipe is screwed tightly into the fitting.

Many jointing compounds are available to help you screw the pipe into the fitting and to seal imperfect or loose threads. But do not rely on these compounds to compensate for poor workmanship. They must not be applied to the female thread as they can be forced inside the fitting when the pipe is screwed and contaminate the water or block a valve.

Figures 3.104 and 3.105 show a cheaper non-adjustable pipe die set. These will produce excellent work and are extremely useful for working in awkward places such as when re-threading pipes which are underground. See Fig. 3.106.

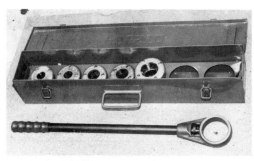

Fig. 3.104 Cheaper pipe die set using non-adjustable dies.

Fig. 3.105 To cut a different size thread change the die.

Fig. 3.106 Threading an underground pipe.

CHAPTER 4

Repairs and Maintenance

TRACTOR FUEL SYSTEMS

Filters

The fuel filter's job is to remove dust and dirt particles which could damage and speed up wear of pump barrels and plungers. It can filter a large volume of fuel without choking or causing a resistance in the fuel flow. There are various types but the most common is the paper element in a canister. The paper strips wound round a cylindrical core in a spiral are welded together at the top and bottom to form a series of continuous V-shaped coils. The paper

Fig. 4.1

Fig. 4.2 Remove the filter by unscrewing the bolt on top of the housing.

used is resin treated and 'creped' for strength (Fig. 4.1).

Most filters are removed by unscrewing a single retaining bolt located on top of the housing (Fig. 4.2). When the filter is removed, note the position of all the sealing rings. (Take great care not to allow any dirt to enter the housing or the new filter.) Some filters include

Fig. 4.3 Some systems have a water trap to collect any condensation in the fuel. Water, being heavier than fuel, accumulates at the bottom of the trap and should be drained off at least once a week. The bowl should be removed periodically and cleaned.

a water trap which requires draining regularly (Fig. 4.3). The lift pump usually has a wire mesh filter which should be cleaned periodically (Fig. 4.4).

If any part is removed, or if the system runs out of fuel, air enters and the system must be vented. Venting points are usually shown in the operator's manual. The system shown in the drawing (Fig. 4.5) would be vented as follows:

(1) Ensure that there is an adequate supply of fuel and that the engine stop control is in the running position.
(2) Loosen union A on top of the filter.
(3) Operate the hand priming lever on the feed pump E until fuel flows free from air bubbles, then tighten the union while the

Fig. 4.4 Fuel lift pump servicing: remove the cover and gauze strainer (if fitted) and clean out any sediment that may have collected. Replace the gauze and cover, then vent the system.

(4) Loosen the inlet pipe B at the injector pump and repeat operation 3.
(5) Loosen the lower vent screw C and repeat operation 3.

(6) Loosen the upper vent screw D and repeat operation 3.
(7) Set the throttle lever to the fully open position and slacken all injector unions.
(8) Turn the engine on the starter motor until fuel squirts from the pipes then tighten the injector unions.
(9) Start the engine. If it runs unevenly slacken one injector union for a few seconds then re-tighten.

Injector change

Fuel injectors should be changed every 600 hours. Failure to do so will result in excessive wear, giving poor engine performance and heavier fuel consumption. Injectors cost only a few pounds for a reconditioned set of four and it is not worth trying to get an extra few hours of use from them.

Ensure that you get the correct replacements. If a reconditioned set is obtained before removal of those to be changed, the dealer will want to know the tractor serial number and the engine serial number.

Strict cleanliness must be observed throughout the operation. Carefully blow, or wash off with paraffin and brush, all dust and dirt from around the injectors. A certain amount of dismantling may be necessary prior to this; certain tractors require removal of the fuel tank (Fig. 4.6).

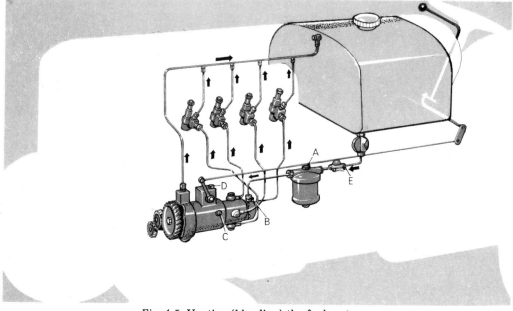

Fig. 4.5 Venting (bleeding) the fuel system.

Fig. 4.6

Turn off the fuel if the tank was not removed in step one and remove leak-off pipes and high-pressure fuel delivery pipes. It may be necessary to drain some fuel from the tank where it has not been removed, as some leak-off pipes are directly connected to the tank and fuel may

Fig. 4.7

leak back. Better still, tackle the job when the tank is nearly empty. Slacken off the pipes at the pump end also, to avoid bending them when removing the pipes from the injectors (Fig. 4.7). Slacken the injector retaining nuts and pull the injectors from their recess. If they are tight use a large open-ended spanner, push one jaw under the lug of the injector and give the other end a sharp blow with the palm of the hand. This should be sufficient to loosen the injector (Fig. 4.8).

Fig. 4.8

Pick out the copper seating washer with a screwdriver or file tang. Clean carbon from the recess with a blunt screwdriver, then turn the engine over on the starter, to blow out any particles from the cylinders.

Fit new copper washer, place the new injector on its retaining studs and tighten the

Fig. 4.9

nuts evenly. If this is not done evenly, the injector will not seat properly and the compression of the engine will blow the fuel past the injector, resulting in poor performance.

Connect up all fuel pipes, making sure all the olives on the pipes are seated properly to obviate leaks. Replace the leak-off pipes, fitting new copper or aluminium washers where required (Fig. 4.9) and turn on the fuel. Bleed the system of air, start the engine and check for leaks.

If diesel fuel affects your hands — it can cause dermatitis — apply a good barrier cream beforehand and wash your hands thoroughly in soapy water when the job has been completed.

Fuel lift pump repair

It should take 10 to 20 minutes to overhaul a pump.

First, remove the inspection cover or glass bowl. Then mark the two halves of the pump, so they can be re-assembled correctly (Fig. 4.10).

Fig. 4.10 Mark both parts of the pump.

Fig. 4.11 Body split to expose the diaphragm.

Fig. 4.12 Methods of retaining the valves.

Remove the screws holding the two body parts together and carefully lift off the top part This will expose the diaphragm (Fig. 4.11).

The diaphragm can be removed by pressing on its centre and rotating it a quarter turn.

The valves may be retained in two different ways (Fig. 4.12).

The valves in the pump body (left) are retained by being 'staked' into position. They are pushed into the body of the pump then a centre punch is used to raise dimples in the sides of the sockets to keep them in. Valves retained this way can be levered out by a screwdriver.

The pump body on the right has its valves secured by a retaining plate which unscrews to remove the valves.

Note which way up each valve is fitted.

When reassembling a pump with a diaphragm spring of different diameters at each end, fit the larger end uppermost.

Refit the diaphragm by inserting the end of its pull-rod into the slot in the rocker arm, then twist it a quarter turn to lock it. Move the rocker arm so that the diaphragm is level with the pump body, then replace the top part of the pump after linking up the marks on the flanges.

After fitting the securing screws, operate the rocker several times so that the diaphragm is fully stretched, then tighten the screws which hold the pump together.

When refitting the pump to the engine, ensure that the rocker arm fits on top of the camshaft and is not inadvertently hooked under it where it could break and damage the engine.

Also check that the arm is not worn (Fig. 4.13) because a worn arm could cause fuel starvation. Figure 4.14 illustrates the contents of a typical fuel pump repair kit.

If the pump is fitted with a glass sediment bowl, do not over tighten it as these are easily broken (Fig. 4.15).

Fig. 4.13 Check that the arm is not worn.

Fig. 4.14 Typical pump repair kit.

Fig. 4.15 Do not over-tighten the glass bowl if fitted.

Safe tank repair

The repair of large fuel tanks is not a job for the non-professional, but small tanks may be soldered, without undue difficulty, after thorough cleaning.

Such tanks may be accidentally dented and though the impact may not cause a leak a rust patch might form and eat through the metal. In these cases, and where a leak has developed as a direct result of mechanical damage, the soldering method can effect a speedy and efficient repair.

First, the tank must be completely cleaned. It should be emptied and inverted to drain before being thoroughly washed out — prefer-

ably with hot water. Allow to dry and check for fuel odours before either washing out again or starting the repair. The damaged area must be scraped and wire-brushed down to bare metal.

Following the sequence illustrated in Figs. 4.16, 4.17 and 4.18 will result in a sound job. Do not attempt to weld or braze a fuel tank of any kind using a naked flame.

Fig. 4.16 After covering the iron tip and damaged area with flux, tin both liberally with solder.

Fig. 4.17 Cut a small piece of sheet steel to cover the damage. Clean it with emery cloth and tin one side.

Afterwards scrub the outside under a hot tap to remove surplus flux; flush the inside with paraffin.

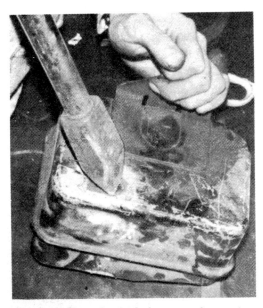

Fig. 4.18 Place the steel plate over the repair and press it with the soldering iron to sweat it on.

TIMING CHECK

To work efficiently, the combustion chamber of a diesel engine must receive the correct amount of fuel, injected at the correct time in relation to the position of the piston. Bad timing can cause excessive fuel consumption, loss of power and premature engine wear.

It can also cause engine 'knock', but few people can discern this noise above the normal rattles of a diesel engine.

A typical injection timing is between $15°$ and $30°$ before top dead-centre (TDC) — when the piston reaches the top of its compression stroke. A figure of $15°$ BTDC (before top-dead-centre) means that the crank-shaft has $15°$ more to rotate before a particular piston reaches TDC.

Many diesel engine flywheels are marked in degrees from 30 BTDC to TDC. Some are only marked at the particular position when injection commences. The flywheel marks can be seen through a small inspection hole in the flywheel housing (Fig. 4.19). The exact injection timing will depend on the design of the engine and can usually be found in the instruction book.

Fig. 4.19 Flywheel marks seen through the inspection hole.

The precise method varies from engine to engine.

The general procedure is as follows: set the piston of number one cylinder to TDC on the compression stroke when both valves will be closed and turn the engine backwards until the injection timing marks line up. These marks can be found on the front pulley of some engines. Rotate the engine back a little, then bring it forward again until the timing marks are once more in line. This will eliminate any backlash in the gear train which drives the fuel pump.

Turning the engine may be difficult. Often the only method is to insert a screwdriver through the flywheel mark inspection hole and lever the flywheel round on the teeth of the starter ring gear.

Some engines can be locked in the position at which injection commences using a steel pin inserted through the flywheel housing and into the flywheel.

There are two common types of fuel pump. The timing of the in-line type, where all the injector pipes are in a line, can be checked once the engine has been set in the correct position by making sure that the timing mark on the drive coupling is lined up with a stationary mark somewhere on the pump body (Fig. 4.20) or the engine valve-drive casing. The coupling is slotted to permit the timing to be adjusted.

Fig. 4.20 Timing marks of an in-line pump.

Modern pumps often have a drive arrangement housed in the valve-drive gear casing at the front of the engine. Timing can be adjusted through an inspection cover by loosening the screws which secure the drive coupling to the drive gear and turning the pump drive slightly while the engine is stationary (Fig. 4.21).

Fig. 4.21 Adjusting the timing of a modern in-line pump.

An accurate check of fuel-pump timing which can be carried out on in-line fuel pumps

is called a spill-timing check. Thoroughly clean the injection pump, remove the high pressure pipe from the pump outlet which supplies number one cylinder and take off the delivery valve housed under the fuel outlet connection. A special tool may be needed (Figs. 4.22, 4.23 and 4.24).

Fig. 4.22 Removing a delivery valve.

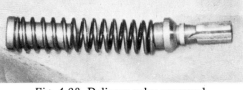

Fig. 4.23 Delivery valve removed.

Fig. 4.24 Delivery valve holder removal tool.

Place the valve assembly in a clean place and replace the delivery-valve holder. Set the piston of number one cylinder on its compression stroke and turn the engine backwards by about 10°. When the fuel lift pump is operated, fuel should flow freely from the delivery-valve holder. Slowly turn the engine forward continuing to operate the lift pump. When the injection timing marks on the flywheel line up, fuel should cease to flow from the delivery-valve holder. This is the point of spill cut-off and indicates the engine position at which the pump would start to inject fuel into the combustion chamber.

Some manuals advise using a swan-neck tube placed on top of the delivery-valve holder to help determine the exact point of spill cut-off. But this is usually not necessary.

Sometimes a spill test will show that a pump is correctly timed, although the timing marks on the pump drive coupling are not exactly lined up. In such cases use the spill test result as being the most accurate check. Do not attempt to change the pump timing unless the correct procedure is certain.

Some fuel pumps have a timing slot on the rear end of the camshaft. This slot should line up with a stationary mark on the pump body at the point of injection.

To check the timing of these pumps set the engines as described above, remove a small inspection cover from the rear of the pump and note whether the slot is in line with the mark on the pump body when the appropriate flywheel timing marks are in line.

Figure 4.25 shows an old injector needle placed in the slot to check that the slot is correctly aligned with the mark in the pump body.

When the timing mark on a rotary fuel pump's mounting flange is lined up with the zero mark on the pump mounting plate (Fig.

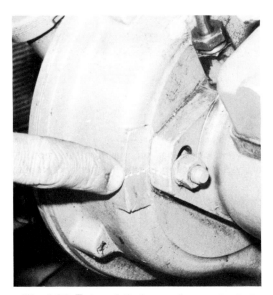

Fig. 4.26 External timing marks on a typical rotary pump.

4.26) the timing should be correct. But not all fuel pumps should be set to the zero mark. Check the workshop manual before changing the pump timing which is carried out by loosening the three pump-securing bolts and rotating the pump slightly.

The timing of some rotary pumps can be checked by first removing the small inspection cover on the side of the pump so that the drive plate of the rotor can be seen.

Fig. 4.25 Using an old injector needle to help check pump timing.

Fig. 4.27 Timing mark on rotor about to line up with circlip edge.

Then turn the engine slowly so that the number one cylinder is on its compression stroke and the flywheel marks are aligned.

A certain letter or number on the rotor drive plate should then line up with a stationary timing mark inside the pump housing. Some fuel pump manufacturers use the edge of one of the ears of a circlip which is inside the pump as a timing mark (Fig. 4.27).

Others discourage anyone from attempting to check the pump in this manner and mount the circlip so that its ears are not visible.

Always clean the outside of the pump and refer to the manufacturer's literature before attempting to check timing in this manner. Never break the seal on a fuel pump in order to look inside it unless instructed to do so by an authorised dealer.

The fuel pump timing of most engines can be checked quite easily. When the timing is incorrect seek professional advice before attempting to change it.

WAXING OF DIESEL FUEL

Many tractors stop or falter during cold winters because fuel pipes and filters become blocked with wax.

The refining process leaves a proportion of wax in diesel fuel. And in freezing temperatures this crystallises and forms lumps. The engine can then fail to start, cut out after a short period, or run on reduced power.

Wax crystals usually form in summer diesel when the temperature drops below freezing point and below minus $9°C$ ($16°F$) in winter diesel.

The main difference between summer and winter diesel is in the refining process. Wax is present in smaller crystal form in the winter type, with a consequent reduction in the temperature at which it starts to congeal.

Some fuels can be refined to winter standard without any further assistance. Others require an additive to help bring them down to the British Standard of minus $9°C$.

Many farmers have requested fuels which give more protection against low temperatures. A spokesman for Shell said that producing diesel to Scandanavian specifications — protection against waxing down to minus $15°C$ ($5°F$) — would reduce the amount of diesel available in Britain by up to 20% because changes in the

refining would be needed. More additive is not sufficient.

To reduce the risk of waxing, make sure fuel storage tanks are not contaminated. Water should be drained regularly and containers cleaned out. Ideally, have one tank for winter fuel and one for summer fuel to prevent dilution by the summer diesel when winter stocks are bought in.

First check that the stoppage is not caused by ice. Fine ice crystals can look like wax. Draw fuel from the tank, filter it into clean, dry bottles and allow it to stand. Water will separate at the bottom as a clear layer or globules.

Where possible keep tractors under cover when temperatures drop and wrap fuel lines and tanks with an insulating material.

Should waxing occur, heat the affected area, but *never* use a naked flame. Use hot water, towels, air or electric heating tapes.

A number of additives are available for left-over summer diesel. They are claimed to reduce the risk of waxing at low temperatures. Cost of treatment is about 1p a gallon.

The additives are claimed to reduce the waxing temperature of summer diesel by $11°C$ ($20°F$).

Mr Roger Hutchinson of Shell said that the additives would not normally make winter fuel any more resistant to waxing.

'On some fuels it may have a small effect and on others none at all,' he said. 'Even some winter diesel supplied without any additive at all will not respond as it has already been refined to the limit of its low-temperature characteristics.'

Mr Hutchinson stressed that extra additive should be poured direct into the vehicle tank. Temperatures should not be below freezing at the time as inadequate mixing could result.

TAPPET ADJUSTMENT

Modern tractors are so complex that when major components like the hydraulics go wrong, a dealer's mechanic has to be called in. But there are many servicing jobs that can be done by the driver. One of these is tappet adjustment.

A valve in an engine is opened by the cam shaft through a push rod which operates a centrally-pivoted rocker arm. As one end of the rocker arm is raised by the push rod, the other end presses down on the valve stem and opens the valve.

Since different metals expand at different rates, there must be a gap between the valve stem and rocker arm to allow the valve to seat properly when hot.

This gap is referred to as valve or tappet clearance. If the clearance is too small there will be loss of power and burning of the valve seats (Fig. 4.28).

Fig. 4.28 Burnt out valves.

Too big a gap will mean a noisy engine, and valves will fail to open properly. Wrong setting either way results in poor engine performance.

First step in adjustment is to rotate the crankshaft until the piston in the cylinder on which the valves are to be adjusted is at top dead centre with both valves closed. You can

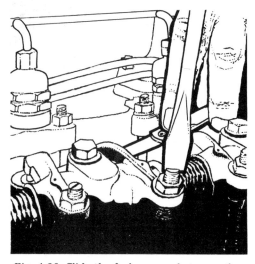

Fig. 4.29 Slide the feeler gauge between the valve stem and rocker arm. Tighten the screw until the feeler is a light dragging fit. Then tighten up the nut while still holding the screw firm.

then measure valve clearances with a feeler gauge and adjust them accordingly (Fig. 4.29).

This method can prove difficult for someone not familiar with engines, and many mechanics prefer the following way.

For a four-cylinder engine, which has two valves to each cylinder and so has eight tappets to adjust, take the number 9 as a guide. Work from the front, and turn the engine until valve 8 is fully open. Then you can adjust the gap on tappet 1 (8 + 1 = 9).

Similarly, to adjust tappet 2, valve 7 must be fully open (2 + 7 = 9). Continue like this until all gaps are correct.

This method is known as the rule of 9, and is applicable to most four cylinder in-line tractor engines, but the valves of some engines cannot be adjusted in this manner. There is also a rule of 13 for six cylinder engines.

Three cylinder engines require a different method of tappet adjustment. A typical method would be to set the engine with number one piston at TDC on the compression stroke and adjust valve numbers 1, 2, 3 and 5 then turn the engine one complete turn and adjust valve numbers 4 and 6. A foolproof method of adjusting the valve stem clearances (tappets) on any four stroke engine is to rotate the engine until a particular valve is open, then rotate the engine one complete revolution, and adjust it; this method is based on the fact that when the crankshaft turns once the camshaft will only turn half a revolution, thus the valve will travel from being fully open to being fully closed, which is the position in which it should be adjusted.

Turning the engine may prove difficult. Often the only method is to insert a screwdriver through the flywheel inspection hole and lever the flywheel around on the teeth of the starter ring gear. In extreme cases the starter must be removed in order to gain access to the flywheel.

COOLING SYSTEM CARE

An hour spent checking the tractor cooling system will help to ensure a troublefree winter. Examining the thermostat, the radiator and the antifreeze are the three main jobs.

The thermostat is a temperature-sensitive valve in the cooling system to allow rapid warm up of the engine and keep it at the correct temperature.

Two main types are in common use, the bellows type filled with a volatile liquid, and the wax-filled thermostat. As the temperature rises the volatile liquid or wax will expand so opening the valve and allowing water to pass from the engine water jacket to the radiator to be cooled for further use.

If a temperature gauge is fitted to your tractor, use it to check the thermostat. Starting from cold, the temperature should build up rapidly and remain constant.

Remove the radiator filler cap after starting the engine from cold and allow it to run at a fast idle. There should be little or no flow of water through the radiator. If there is a flow remove the thermostat to check if it is stuck in the open position. (See Fig. 4.30.)

Fig. 4.30 The valve on the thermostat on the left is stuck open and will not allow the engine to reach its proper working temperature.

If you have no temperature gauge, check the thermostat by removing it from the engine and placing it in a beaker of cold water. Heat it on a cooker and use a thermometer to check at what temperature it begins to open. The correct temperature at which it should open will be stamped on the top. It should start to open within 2 or 3°C of the marked temperature and go on to the fully open position within a further 11 to 14°C (see Fig. 4.31).

Check the seals in the radiator cap are in good order (Fig. 4.32). The radiator cap can be tested by your dealer, who will have a special pump which can be used to test both the operation of the cap (Fig. 4.33), and test the cooling system for leaks by pressurising it (Fig. 4.34).

After working in harvest fields with plenty of dust and chaff about, the radiator core will soon become blocked with debris as the fan

draws air in. Use an airline to blow dirt out from the fan side (Fig. 4.35).

Fig. 4.31 Check that thermostat opens at the right temperature by testing with a thermometer in hot water.

Fig. 4.32 Check that the vacuum valve in the filler cap is not jammed by dirt and that the rubber gasket on the cap is in good condition.

Open the tap, drain the radiator and back flush it. Do this by removing the radiator cap and bottom hose. Place one end of a hose pipe into the bottom outlet of the radiator, seal the join with some rag. Connect the other end of the hose to the mains water supply. The pressure of the water will move all the bits of rust and dirt and clean out the down tubes. Continue until clean water comes out of the filler cap.

Lastly make sure that the overflow pipe is

Fig. 4.33 Testing a radiator cap.

Fig. 4.34 Cooling system pressure test.

Fig. 4.35 Use an air line to blow out the dirt from the fan side of the radiator — not from the front.

and in most cases the damaged parts need renewing.

It is possible to drain the tractor of water each night but it is a tiresome chore that can easily be forgotten. Put antifreeze in at the beginning of the winter and top up as necessary. Always use the same type and strength of antifreeze as was originally used. Mixing two types may result in a chemical which damages the metal.

Use clean containers to mix antifreeze and water. Refill with the correct mixture until it is an inch from the top. Run tractor for 10 mins and then check level (Fig. 4.36).

Fig. 4.36 Carefully measure antifreeze in clean containers and mix it with clean water.

not blocked and that the filler cap seal is in good condition.

When mixed with water antifreeze lowers the freezing point. A 25% solution — one litre of antifreeze to three litres of water — is enough to prevent freezing in most UK winter conditions. If in doubt consult your dealer on the correct amount to mix with the water.

Without antifreeze, when water reaches freezing point the ice will expand. Since the cooling system is sealed it will push against the metal of the cylinder block, head and radiator until it finds a weak point, and splits the metal. Frost cracks on an engine are expensive to repair

Check hosepipes for cracks or splits, renew if necessary and make sure all hose clips are tight. Correctly tension the fanbelt and inspect the fan for missing or bent blades. A fan with a blade missing can damage water pump bearings.

FAN BELT TENSION

Most fan belts are tensioned by swinging out the dynamo or the alternator, which usually requires a slightly tighter belt.

Some tractors have a split pulley system. Either the fan pulley or the crankshaft pulley is made in two parts. With these, the belt is tensioned by moving the two halves of the pulley closer together, so squeezing the belt to the outside.

Sometimes the two halves of the pulley are brought closer together by removing shims. More often, the arrangement is similar to the International Harvester type (Fig. 4.37). Here the inner part of the pulley is fixed to the hub, while the outer part is screwed onto the hub. The outer part here is unscrewed completely to expose the threads.

Fig. 4.37 Split pulley type of adjustment.

To tighten such a belt system, loosen the locknut and remove the grub screw from the outer part of the pulley, then screw this outer part along the hub until the belt is sufficiently tight. The grub screw should be inserted only when its hole lines up with one of the slots in the hub or the threads on the hub will be damaged.

LUBRICATION

After use, oil becomes contaminated with dust, carbon deposits, sludge, unburnt fuel, swarf and acids formed from exhaust gases which escape past the piston rings.

Heat tends to oxidise the oil. Contrary to popular opinion, oil must be changed at the recommended period even when it still feels oily. Usually diesel engine oil should be changed every 200 hours, but engines which are used for a lot of stop-start work should have their oil changed more often.

Turbo-charged engines require more frequent oil changes — usually at about 100 hours. The oil filter is usually changed at every second oil change, but many manufacturers now recommend that the filter should be replaced at every oil change.

Before changing the oil and filter, clean the area around the drain plug and filter. Always drain the oil when the engine is hot so that any sludge will be flushed out suspended in the oil. Leaving the engine to drain overnight can cause the oil pump to lose its prime.

Replace the filter with the correct type. This is especially important with a screw-on filter mounted upside down which will probably be fitted with an anti-drain valve to prevent it from emptying when the engine stops.

When replacing cartridge filters check that the plate-sealing washers and spring are located correctly, or the oil may be able to by-pass the filter instead of being forced through it.

When possible fill the filter bowl with oil before replacing it, to reduce the time that the engine has to run while the oil is pumped into the bowl.

The rubber sealing ring in the filter housing is best removed by a sharp spike, such as a safety pin. Trying to lever out the rubber ring with a screwdriver may result in a broken housing.

Screw-on filters are quick to replace when fairly accessible. Special tools are available for the job, but any chain wrench should suffice. Difficult cases can be removed by driving a punch in one side and out the other and using it as a handle. But care must be taken not to bend the mounting screw thread or it will be impossible to make the new filter seal properly.

After changing any oil filter, run the engine slowly until the oil pressure has built up.

Many engines have a pressure gauge in the oil gallery to record the pressure at which the oil is being forced around the engine. Often, instead of a gauge, a switch unit fitted in the oil gallery turns on the oil warning light should the oil pressure drop below 15 psi.

When pressure drops the engine must be

stopped immediately before serious damage is caused.

Common causes of low or loss of oil pressure are:

● *Low oil level in the sump.*
Remedy: check and refill sump.

● *Oil too thin — possibly due to an overheated engine.*
Remedy: check type and condition of oil and the engine temperature.

● *Faulty relief valve.*
Remedy: locate relief valve and inspect spring and plunger assembly. Check the free length of the spring. In some, the pressure can often be increased by adding shims under the spring or removing shims from under the spring retaining cap. Seek professional advice before doing this. The relief valve is often fitted adjacent to the pump, therefore to gain access to it will involve removing the sump. Figure 4.38 shows a typical relief valve assembly. Before removing the sump, ensure that the light unit or the gauge is not faulty by removing the unit and measuring oil pressure (Fig. 4.39).

Fig. 4.38 Typical relief valve assembly.

Fig. 4.39 Measuring the oil pressure.

● *Suction gauze choked.*
Remedy: remove inspection plate or sump and clean the filter.

● *Internal oil leak.*
Remedy: remove sump and inspect all gaskets, seals and pipes connected with the lubrication system. Also inspect the pump for cracks.

● *Crankshaft trouble.*
Remedy: overhaul bottom end of the engine.

● *Faulty pump.*
Remedy: check or replace the pump as follows:

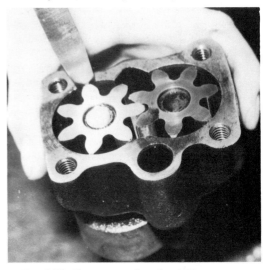

Fig. 4.40 Gears to casing should be no more than 0.1 mm.

Figure 4.40 shows a gear type of pump being checked. Remove the end plate and inspect the gears and the body for score marks or other damage, then check the following clearances — gears to casing, no more than 0.004 in (0.1 mm) gears to end plate, no more than 0.008 in (0.2 mm) (Figure 4.41).

Fig. 4.41 Gears to end plate — no more than 0.2 mm.

Fig. 4.42 Outer rotor to body — no more than 0.2 mm.

Figure 4.42 shows a rotor type pump being checked. To do this, remove the end plate and inspect it for score marks, then check the following clearances — inner rotor to outer rotor, no more than 0.010 in (0.25 mm); outer rotor to body, no more than 0.008 in (0.2 mm), rotors to end plate, no more than 0.004 in (0.1 mm) (Fig. 4.43). These clearances may vary according to make.

Occasionally an engine will have an excessively high oil pressure. This is quite normal when the engine is cold, but if it persists when the engine is warm, investigate it. Excessive oil pressure could burst an internal seal or the pressure gauge. The usual causes of high pressure are incorrect oil, faulty relief valve or a blocked oil passage on the pressure side.

AIR CLEANERS

Two main types of air cleaner are used on agricultural vehicles:

The oil bath type Air is drawn down the centre tube and forced through a 180° turn. This throws out by centrifugal force many dust and dirt particles which are then trapped in the oil at the bottom of the cleaner. Also the air moving over the surface of the oil tends to draw up some of the oil into the gauze, through which the air travels before leaving the cleaner. This helps to filter out any remaining dust particles.

The oil should be changed regularly, depending on conditions — daily when it is dusty. In general, change the oil when the layer of dirt in the bottom of the oil bath exceeds 6 mm (¼ in), or when the oil no longer feels sticky.

Always fill the oil bath to the correct level and with the correct engine oil. An overfull oil bath or one with oil that is too thin may cause the engine to suck the oil out of the air cleaner. In a diesel engine, there is a risk of the engine being damaged due to over-revving. Diesel engines can, in certain circumstances, use the oil from their air cleaners as fuel.

Once this has started to happen, little can be done to avoid engine damage.

When an air cleaner is allowed to operate for long periods without an oil change, the quantity of dirt settling in the bath may be sufficient to displace the oil upwards creating the same effect as over-filling. This is a more

Fig. 4.43 Rotors to end plate — no more than 0.1 mm.

Fig. 4.44 Oil bath type air cleaner.

Fig. 4.45 Dry type air cleaner.

common cause of engines over-revving than an air cleaner being over-filled. Some air cleaners are designed so that they cannot be overfilled.

Occasionally the centre pipe and fixed gauze should be cleaned.

Dry air cleaners These are widely used on farm equipment. Usually air is drawn through a spiral — which removes most of the dirt particles — then through a fine paper element. Many are fitted with an automatic unloader valve — a rubber valve at the bottom of the cleaner which is sucked shut when the engine is running, but opens when the engine stops, to allow the collected dirt to drop out.

Inspect the cleaner regularly. Most tractors now have a device which automatically indicates when the filter requires attention. This works by sensing the extra suction required to draw air through the filter element.

Dry air cleaner elements can be cleaned in four ways:
• Lightly tap to remove loose dry dust from the outside. This is usually the only maintenance required.
• Blow off dust with compressed air directed from the inside to the outside at a pressure not exceeding 2 bar (30 psi).
• Hose off the dust in much the same way as with compressed air — from the inside.
• Wash the element in a sudless detergent and allow to dry.

Never clean one of these elements by washing it in diesel oil.

After washing a dry paper element, inspect it for punctures in a dark place by putting a light inside and checking for holes.

The hoses connecting the air cleaner to the inlet manifold should be checked. It is pointless having a well-serviced air cleaner and then allowing dirty air to be sucked in through a cracked hose — this applies particularly to turbo-charged engines. Particles of grit landing on the blades of a turbo-charger can cause a lot of damage.

Pre-cleaners are used to remove coarser dirt particles before they enter the air cleaner. The air swirls round inside throwing the heavier particles to the outside where they are ejected through slots or slide down the outside to be caught in a bowl.

Do not operate an engine without its pre-cleaner unless absolutely necessary.

COMPRESSION

A compression test on an engine can determine whether a full engine or a top-end overhaul is necessary.

Low compression causes deterioration in the performance of petrol and diesel engines. A diesel engine relies upon cylinder compression to create enough heat to ignite the fuel, too low a compression and the engine may not start.

The compression depends upon the compression ratio of the engine — a comparison of the volume of the cylinder and combustion chamber when the piston is at the bottom of its stroke, with the volume remaining when the piston is at the top.

A diesel engine has a much higher compression ratio than the petrol engine. Pressure at the end of the compression stroke will vary with engine design, generally, 8 bar to 11 bar for a petrol engine and 25 bar to 30 bar for a diesel. A small two-stroke engine on a chain saw would be about 6 bar to 7 bar. Due to the variation between different engines, the correct figure must be checked with the manufacturer's specification. Modern workshop manuals often state pressures in atmospheres or bars. Both signify how many times the pressure concerned is greater or smaller than atmospheric pressure.

To change into psi, multiply by 15; the answer will be accurate enough for a compression test.

Testing a diesel engine

Figure 4.46 shows the test kit. This consists of a pressure gauge, reading up to 50 bar, the adaptor, which replaces the injector in the cylinder being tested and contains a valve which allows the gauge to be pumped up to the maximum pressure instead of flicking. These test kits are not expensive in relation to the work they can save.

Fig. 4.46 Typical gauge and adaptor used for compression testing.

Fig. 4.47 Compression gauge in position.

The adaptor must be correct for the engine. Clean the area around the injectors so no dirt can fall into the engine when they are removed. Run the engine to working temperature then stop it and remove the injectors and their copper sealing washers. Carefully clean all the injector holes and insert the adaptor for the compression gauge into number one cylinder injector hole, using a good copper washer to seal it and connect the gauge unit (Fig. 4.47).

The gauge can usually be fitted using a steel or a flexible pipe. Pull the stop control fully out and use the starter to crank the engine at least at normal cranking speed — normally 200 r.p.m. to 300 r.p.m. Engines with pneumatic governors must have the throttle open but very few modern engines have such a governor.

Continue to turn the engine until no further pressure rise is recorded on the gauge, which should hold maximum reading until it is released by the button.

Note maximum pressure for number one cylinder and repeat the process for the remaining cylinders.

To save wear and tear on the starter motor and battery do not completely release the pressure in the gauge unit when moving it from cylinder to cylinder. This will avoid having to pump the unit up from zero each time. Normally the pressure need be reduced only to about 14 bar between readings.

Maximum pressure readings should be within 3 bar to 4 bar of the specified value. More important, the readings for each cylinder should be within 1.3 bar to 1.6 bar of each other. A cylinder pressure of more than 1.6 bar below the other cylinders indicates a gas leak.

Should the test indicate that two adjacent cylinders have a low compression, the head gasket should be suspected before the valves and rings.

Where the pressure readings are higher than the specified value, this can be caused by excessive carbon build-up in the cylinder head.

Low compression pressures should be retested after a tablespoon of engine oil has been squirted into the appropriate cylinder, which should help to seal the piston rings. Where the second test shows an improved compression, the problem is probably worn rings. Should the addition of the oil produce little or no improvement, the loss of compression is likely to be due to valve trouble. Serious valve trouble should also be suspected when the pressure is very low or fails to rise at all.

Testing petrol engines

A compression test on a petrol engine is simpler because the gauge adaptor is pushed into the sparking plug hole. This gauge can be removed and fitted to a different adaptor to measure oil pressure.

This unit also has a non-return valve which allows the gauge to hold the maximum pressure until the pressure release button is pressed (Fig. 4.48).

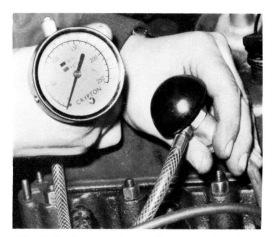

Fig. 4.49 Compression testing a petrol engine.

TOP END OVERHAUL

Apart from big overhauls, the cylinder heads of tractors used for field work will normally only be removed to recondition or replace burnt valves, or a blown head gasket. It is one of the few tractor-repair jobs that does not need much expensive equipment.

Figure 4.50 shows a typical set of the necessary equipment. This includes a valve spring compressor, valvegrinding stick and paste, and a torque wrench. When special treatment is needed, consult a professional engine-servicing firm.

Before removing a cylinder head, read the relevant manual carefully to check whether it is

Fig. 4.48 Compression gauge adaptor for petrol engines.

The adaptor is pushed into the plug hole (Fig. 4.49) and the engine cranked on the starter. For petrol engines the throttle must be wide open so that the engine's breathing is not restricted. The results can be interpreted in the same way as the diesel engine test except that the pressures should be within 10% of the specification.

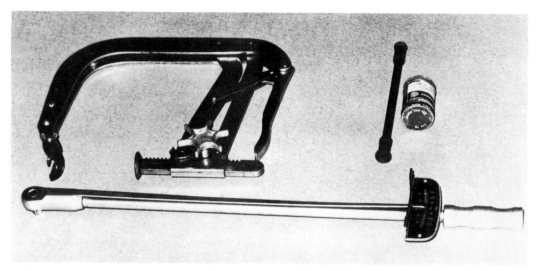

Fig. 4.50 Equipment required to decoke an engine — valve grinding stick and paste, valve spring compressor and a torque wrench.

necessary to do so. The amount of carbon build-up can often be judged by removing the inlet and exhaust manifolds and looking into the ports. Lack of compression may be due to incorrectly set valve clearances or a badly blocked air cleaner.

Most of the preliminary dismantling necessary to get at the cylinder head is obvious, but the following points should be noted.

First, remove the bonnet and thoroughly clean the engine. While the engine is drying, drain the cooling system, including the block, and remove the battery. The battery should be cleaned, and put on charge at a low rate.

When necessary, remove the fuel tank and such items as fuel and leak-off pipes. The exhaust and inlet manifolds usually come off at this stage. The open ends of injector pipes should be covered unless they are to be replaced and the unions of the injectors should be blanked off to prevent dirt getting in.

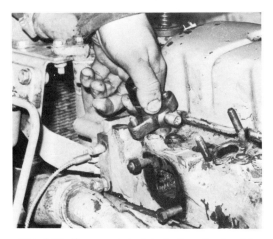

Fig. 4.51 Remove injectors before the head is taken off to avoid any possibility of damage.

Remove the injectors before the head is taken off to avoid any possibility of damage (Fig. 4.51). This applies particularly to direct-injection engines where the tips of the injector protrude slightly from the cylinder head. This is the case in most Ford engines.

Removing the rocker cover allows access to the rocker shaft which usually comes off complete with the rockers after releasing the securing bolts. Often there is a small oil pipe connecting this shaft to the head which must also be disconnected.

Keep the push rods separate unless they are identical. Do not pull up the cam followers

with the push rods as they may be difficult to relocate and do not drop a push rod. In some engines, such as the M-F 175, the push rod can drop down the tube into the sump.

Sometimes the nuts holding the rocker shaft also secure the cylinder head. When this is the case, loosen the head-securing nuts at the same time as these nuts. All head nuts should be loosened one turn at a time and in the correct sequence. This prevents distortion. Caps fitted on top of the valve stems must be removed (Fig. 4.52).

Look out for special screws or nuts which, when they are unscrewed, will lift the head. Such screws or nuts must not be unscrewed until all the other nuts have been removed.

Should the head prove difficult to lift, use lugs on the head as lever points. Do not force up the head by wedging a screwdriver into the gasket.

Fig. 4.52 Caps fitted on top of the valve stems must be removed.

Sometimes the head becomes stuck to its studs. Try to remove the offending stud by locking two nuts together on its upper section and then unscrew it from the block.

Check that the head is not being restrained by a hidden nut, stud or by-pass hose. Try carefully lifting a tight head with a hoist fastened to plates drilled to fit over the manifold studs.

With the head off, inspect the head gasket. Oil in the water or water in the oil are tell-tale signs that a gasket has blown. So are overheating and loss of power.

Sometimes the engine may appear to be seized when the starter key is turned. This is due to water finding its way into the combustion chambers. Any engine displaying such symptoms should have its gasket carefully inspected.

Figure 4.53 shows how small the rupture can be. Gaskets can blow for no apparent reason, but usually the problem is faulty fitting, loose head bolts, warped head or dropped cylinder liners.

Fig. 4.53 The pointer indicates a small rupture in the head gasket.

Fig. 4.54 Washers being used to retain liners while the head is off.

Unless it is certain that the engine is not going to be turned while the head is off, the cylinder liners should be secured by slave washers and screws or nuts (Fig. 4.54). Otherwise there is a risk that the liners will lift when the engine is turned.

Before removing the valves, clean them with a wire brush to check that they are not already marked, otherwise use a centre-punch to number each one starting from the front or water-pump end of the engine (Fig. 4.55).

The valves are removed with a valve-spring compressor (Fig. 4.56). Should the compressor

Fig. 4.55 Marking the valves with a centre-punch.

Fig. 4.56 Removing the valves with a valve spring compressor.

fail to operate, the collets may be jammed in the spring retainer.

Apply a little force to the compressor, then strike the end of the compressor or spring retainer (Fig. 4.57) which should free the collets.

Fig. 4.57 Freeing stuck collets. Apply a little force to the compressor, then strike the end of the compressor.

During valve removal, note the position of any rubber seals and whether any spacers are fitted under the springs. Look for differences in the springs. In some engines the exhaust valve spring is heavier.

Next, clean the valves. Put them in the chuck of an electric drill and use a broken hacksaw blade to remove the worst of the carbon, then finish the cleaning with emery paper (Fig. 4.58). The inlet and exhaust ports of the head must be thoroughly cleaned using a scraper and wire brush.

Fig. 4.59 Cleaning the valve ports.

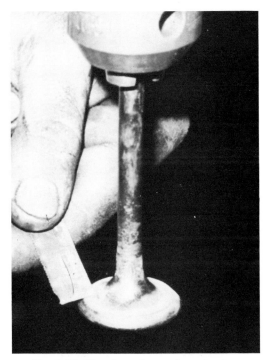

Fig. 4.58 Cleaning valves using a broken hacksaw blade.

A wire brush in an electric drill is useful (Fig. 4.59).

The flat surface of the head must be carefully cleaned with a soft scraper and wire brush. This surface and the mating surface on the top of the cylinder block must be made perfectly clean or the new gasket will not seal correctly.

Do not remove the pre-combustion chambers on indirect injection engines at this time. A perfect seal must be retained between the pre-combustion chamber and the head. Should you have to remove these chambers, make sure that all surfaces are scrupulously clean and all relevant gaskets are in place when the chamber is refitted. With the head free from all carbon, wash it in paraffin or a degreaser. Diesel oil can

be used, but it does not have the same cleaning action. Rinse with water.

While the head is drying, clean the tops of the pistons. Do not scratch aluminium pistons and prevent carbon from falling down the push rod holes or into the water ways. Some manufacturers advise leaving a small ring of carbon around the outside of the piston crown. They claim this reduces the risk of increasing oil consumption.

Check that the cylinders do not have excessive lip where the piston-ring travel ends. This indicates that the cylinders are becoming worn and will soon require replacing or reboring.

Check that the water ways are clear and that there are no cracks in the block. A cracked block may well have caused the same overheating symptoms as a blown head gasket.

Where applicable, check that the cylinder liner protrudes above the cylinder by the amount stated in the workshop manual — usually 0.1 mm to 0.15 mm. A dropped liner will cause the head gasket to blow. This requires professional help as the liners must be removed.

Check the cylinder head for cracks on its flat surface and across the valve seats. Place each valve in its correct seat. Then lift each one slightly (Fig. 4.60) and try to move it from side to side.

Should the movement be more than 0.1 mm, the valve guides and possibly the valves should

Fig. 4.60 Checking the valve and guide for wear.

be replaced. The guides are extremely brittle but can usually be tapped out using a specially made tool. The new guides can often be tapped back in.

The guides will be easier to fit after chilling in a deep freeze for two hours before fitting and the head heated evenly until it is just bearable to touch.

Make sure that the new guides are inserted the right way up and that they project into the ports by the correct amount.

Sometimes guides can be pulled into place using a long bolt and a series of suitable washers. But this method takes so long that it is no use heating the head and cooling the guides because they are back to the same temperature before the job is finished.

Fig. 4.61 Valve with hollow face.

After dealing with the guides, check that the valve faces are not hollow or burnt (Fig. 4.61). Hollow valves must be refaced before they can be lapped in. This requires a valve-facing machine which is a specialist piece of equipment.

Having the valves refaced professionally is much cheaper than purchasing a new set. Have the seats cut at the same time. The specialist can check the valve and seat dimensions and reduce the valve seat width where necessary. This is essential to get high seat-pressure.

Finally, when checking the valve train for wear, inspect the faces of the rocker arms for deep wear marks. Worn rockers should also be refaced professionally.

Assuming the head and valves are in reasonable condition, which is normal, the next step is to lap each valve to its seat to form a gastight seal.

Fig. 4.62 Lightly oil the valve stem.

Oil the valve stem with a light engine oil (Fig. 4.62); place a small quantity of coarse grinding paste on the valve seat in several places; place the valve on its seat, and rotate by about half a turn in either direction.

Attachments are obtainable for electric drills. These will speed up the job, but are not essential.

Figure 4.63 shows a valve being ground with a grinding stick which has a rubber sucker on its base.

Some valves on smaller and older engines have slots so that a screwdriver can rotate them. The valve must be raised at intervals and lowered into a new position to spread the grinding paste

Fig. 4.63 Grinding a valve using a grinding stick.

Fig. 4.65 Valve spring with close coils at one end.

and obtain a more even lapping action.

Continue grinding until both the seat and the valve face have a continuous dull grey finish.

Normally, the seat will take longer than the valve but do not continue grinding more than necessary.

To test the seal, draw several pencil lines across the seating in the head then replace the valve and turn it completely. If each line is cut, the valve is seating properly.

Each valve should be given a final lapping with a fine grinding paste.

Remove every trace of grinding compound from valves and head. When necessary the head must be re-washed. At this point, remove the thermostat and test it.

Valve springs should be tested for distortion or weakness. Ascertain its length by measuring it against a new spring or comparing it with the others. Springs can be tested under load in a vice end-to-end with the new one (Fig. 4.64). When the older spring has weakened, it will be more compressed than the new one.

Some engines have springs with closer-spaced coils at one end (Fig. 4.65). These must be fitted so that the closely wound coils are next to the cylinder head.

Fig. 4.66 Positioning a valve seal.

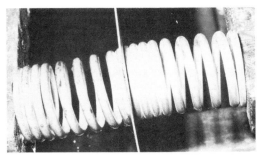

Fig. 4.64 Checking a valve spring.

Oil the stem of one valve, place it in its original seat, fit a new oil seal (Fig. 4.66) and refit the spring and retaining assembly using a spring-compressor. Check that the spring-

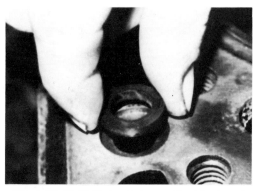

Fig. 4.67 Refit O-ring between the block and the head.

retainer is located properly by covering the spring with a cloth and striking the valve stem several times with a hammer.

Before replacing the head, remember to fit the O-ring between the cylinder block and head where necessary (Fig. 4.67). Also check that the mating-surfaces are scrupulously clean and ensure that the head gasket is fitted the correct way up.

Gaskets are often marked top and front. The composite type is normally fitted copper surface uppermost.

Unless otherwise stated by the manufacturer, do not use a jointing compound on the gasket because this will do more harm than good. Many gaskets are coated with their own special lacquer which must not be removed before fitting.

When the head is retained by bolts, use slave-bolts to locate the gasket. These are then removed after the head is secured (Fig. 4.68).

Use a torque wrench to tighten the cylinder head in several stages and in the correct sequence to prevent the head becoming warped or cracked during the process.

The head should be re-tightened after the engine has been run for a few hours. The gasket is likely to become more compressed after a few hours' use and the head securing bolts may lose tension. This could lead to the head gasket blowing again. It is more important to re-torque short studs than long ones.

Refit the push rods and rocker shaft, including any small oil pipes that may be connected to the rocker shaft. Before refitting the injectors,

Fig. 4.68 Slave-bolts retaining the gasket while the head is fitted.

set the valve clearances because, with the injectors out, the engine will be easier to turn.

Having set the valve clearances, refit the injectors or an exchange set where they have been in service for more than 500 hours. Otherwise, test the original set for correct spray pattern and cracking pressure.

Ensure that the holes in which they sit are absolutely clean and that a new copper sealing washer is used. When new washers are not available, anneal the old ones by heating to cherry-red and quenching in water.

Install the leak-off pipes, remembering to fit sealing washers above and below the banjo unions. Connect the injector pipes but leave the unions slack on the injector end.

The oil and filter should be changed when the engine oil is contaminated by water.

Pour a little oil over the rockers and push rods for initial lubrication.

Finally, replace the rocker cover and all the ancillaries, and refill the radiator. Replace the battery and crank the engine with the starter to bleed air from the fuel system and tighten the injector unions.

Check that any starting aid which may be fitted is connected properly. This is important in engines employing glow-plug heaters where it is easy to create a short.

With the engine running at about 1000 r.p.m. for five minutes, check for leaks and look into the rocker cover to see that oil is reaching the rockers and valves.

REPAIRING RECOILS

Several types of starter systems are used on small engines and the method of transferring the torque from the starter pulley to the engine flywheel will vary from engine to engine.

The engine discussed in this section is a 3 hp Briggs and Stratton. It uses a spiral spring to recoil the starter cord after it has been pulled. To remedy difficulties, such as a faulty over-run device and a broken cord or recoil spring, proceed as follows.

Fig. 4.69 Removing the rewind housing.

Fig. 4.70 Unscrewing the overrun device.

First, remove the rewind housing (Fig. 4.69). If the over-run device is giving trouble it must be unscrewed from the crankshaft (Fig. 4.70). The over-run may need a sharp tap with a soft-headed hammer to loosen it. On some engines the over-run is threaded left-handed to prevent it becoming unscrewed when the starter rope is pulled.

Dismantle the over-run and check that the ball and ratchet mechanism is not damaged or stuck. The easiest way to reassemble is to insert the ball bearings after the centre is in place (Fig. 4.71).

Fig. 4.71 Assembling the overrun device.

To check the recoil mechanism, bend up the tabs which retain the cord pulley and recoil spring (Fig. 4.72) and remove the pulley. Insert

Fig. 4.72 Bend back the tabs which secure the pulley and spring.

Fig. 4.73 Tie a knot in the end of the new cord.

a new cord if required through the hole in the pulley and tie a knot in the end (Fig. 4.73).

The recoil spring usually breaks at the end, and can often be re-used by carefully bending the end into a new hook. In the starter shown, it is easy to regrind the end of the spring to the correct shape for refitting.

To reassemble the recoil starter, insert the cord into the pulley and through its hole in the housing, but do not coil it up. Coil up the spring and fit it into the housing so that the end is correctly located in the slot in the housing (Fig. 4.74).

Fig. 4.74 The first step is to place the spring in the housing.

Lightly oil the spring, then carefully lift the middle end, attach it to the pulley and replace the pulley in the housing. This is easier when the pulley is slightly rotated so that the spring is coiled tighter.

Having located the spring and pulley, tension the spring so that it can rewind the cord. Pull the cord out from between the outside of the pulley and the inside of the housing (Fig. 4.75). Many pulleys have a slot into which the cord fits. Rotate the pulley about three or four times in the same direction as the cord runs.

Fig. 4.75 Pull the cord out from between the pulley and the housing.

Carefully release the pulley and check that it completely rewinds its cord. Should the cord not fully recoil, tighten the spring by rotating the pulley one or two extra turns using the method described above.

Finally, secure the pulley in the housing by bending over the retaining lugs (Fig. 4.72). Tightening the spring this way is much easier than attempting to wind up the rope before fitting the pulley into the housing.

When fitting a new recoil spring or cord, the spring should only be pre-tensioned sufficiently to fully rewind the cord, otherwise it will be over-strained. And the cord must be short enough to limit how many times the pulley rotates when the cord is pulled to the fully extended position. The rotation of the pulley must *not* be limited by the spring.

Even when the spring and cord are in good condition, the cord can still be hard to pull and may only partially return when the handle is released. The mechanism is probably clogged with dirt. This will be more apparent at low temperatures because oil in the recoil spring can thicken enough to stick the coils together. The spring should be removed, cleaned and lightly oiled before replacing.

Some starters stick or even fail to engage due to damaged pawls (Fig. 4.76). Remove the starter housing and check the pawls for wear or sticking.

Fig. 4.78 Check for movement of the king pin.

Fig. 4.76 Some starters are fitted with pawls which may become damaged or stuck.

WHEEL BEARINGS

Most front-wheel bearings, especially on tractors, are of the taper roller type and are adjustable to compensate for slight wear.

To check the bearings, jack-up the wheel and grasp it firmly at 'six o'clock'. Do not place fingers under the wheel unless the axle is on a support stand. Attempt to pull the top of the wheel outwards and push the bottom of the

Fig. 4.77 Checking the wheel bearings.

wheel inwards at the same time, then reverse the process (Fig. 4.77).

Ignore any movements between the king pin and its bushes. Keep an eye on the top and bottom of the king pin during this test.

Figure 4.78 illustrates where to look for king pin wear — the wheel has been removed for clarity. Should little or no play in the wheel bearing be evident, spin the wheel, check freedom of movement and listen for any unusual noises such as rumbles, knocks or squeaks.

To adjust the wheel bearings, first clean the area around the hub-cap before removing it. The cap usually unscrews, but the help of a large pipe wrench may be required.

Remove the split pin which retains the adjusting nut by straightening and pulling it out with side-cutting pliers. Tighten or slacken the nut while the wheel is rotated.

Most manufacturers recommend that the nut is tightened until the wheel becomes fairly stiff to turn and then slackened until it just rotates freely.

When the split pin cannot be inserted unscrew the nut until it can.

Some adjusting nuts on the tractor's left side have left-hand threads as a safety precaution. Ideally a new split pin should be used. When the split pin is opened to retain it, make sure that neither of its 'legs' protrude into the bearings.

Some manufacturers recommend tightening the adjusting nut to a specified torque before loosening it again. Others even state a torque to which the nut must be tightened and then left at this torque. In such cases use a torque wrench. Do not attempt to guess.

When a bearing requires replacing, first loosen the wheel nuts, jack-up, place a strong stand under the front axle and remove the wheel.

Remove the hub cap and split pin, and unscrew the adjusting nut. Remove the thrust washer under the adjusting nut, shake the hub and pull it outwards so that the outside bearing is dislodged from the stub axle. Usually the thrust washer and the outside bearing can be removed together.

The hub can now be carefully pulled off the stub axle. Catch the inside bearing if it falls out of the hub. Sometimes the inside bearing will remain on the stub axle and will have to be carefully levered off using two strong screwdrivers.

When the inside bearing is to be reused, it can be left on the stub axle and cleaned in place. Usually, the seal in the hub behind the inner bearing must also be removed unless the inside bearing is to be left on the stub axle (Fig. 4.79).

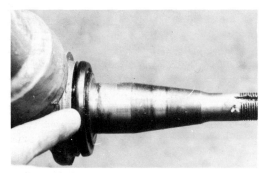

Fig. 4.79 Grease seal position.

The most difficult part is usually the removal of the two remaining parts of the bearing — the outer races of the outside and inside bearings — from the hub. Clean off all grease from the inside of the hub before examining it to find out whether it has been fitted with two slots in the shoulder against which the bearings fit. If so then the outer races can be driven out with a punch in these slots.

Bearings without slots can usually be removed with a socket that will just fit through the bearing. It can then be jammed across to one side with a screwdriver so that it catches on one edge of the bearing.

Before replacing the bearings, examine the hub for signs of burring or other damage to the bearing seats and carefully file down any high spots.

The new outer races should be carefully pressed or tapped into the hub. A useful tool can be made from an old race ground down slightly to prevent it sticking.

When it is necessary to tap the outer race into place using a tool made from an old bearing, avoid using a steel hammer to strike the old race or it may splinter. Always use a nylon hammer or place a block of wood between the old race and the hammer.

It is usually easier to fit the inside roller assembly on to the stub axle instead of the hub. This can easily be pushed into place with the aid of a piece of pipe (Fig. 4.80). Place the seal on to the axle before the bearing, unless it can be fitted later.

Fig. 4.80 Fitting roller assembly.

Before refitting the hub, ensure that the bearing is well greased. Push grease between the rollers. The hub should only be partially filled with grease otherwise grease may be forced past the seal should pressure build up due to heat and churning in use.

After fitting the hub, the outer bearing assembly and thrust washers can be refitted and the bearing adjusted.

BRAKES

Brake adjustment

First step is to jack up the tractor wheel (Fig. 4.81). Tighten the brakes, using a screwdriver through the adjustment slot and apply the brakes hard. This will centralise the shoes and ensure the whole area is in contact with the brake drum, so that it will come away from the drum evenly when it is slackened off (Fig. 4.82).

Fig. 4.81 Jack up the tractor wheel at a convenient point on the axle.

Fig. 4.82 Locate the brake adjuster, tighten fully with a screwdriver, and slacken until the wheel revolves freely.

Slacken the adjuster until the wheel will revolve freely without the shoe rubbing on the drum. Carry out the same procedure on the other wheel.

Road-test on a hard, level stretch of road. Lock both independent brake pedals together then drive forwards in a low gear at about half to three-quarter throttle. Apply the brakes, while holding the steering wheel lightly, and

note which way the tractor veers. Do not depress the clutch pedal during this test. If the tractor veers to one side, the brake should be loosened on that side.

When the brakes appear to be balanced it is safe to depress the clutch while the brakes are applied and the tractor is moving at about 10 kmph. Note which wheel stops first. When the job is complete, make sure the rubber bung or metal plate is replaced in the adjusting slot to prevent water or dirt entering the brake.

Adjust your brakes often if one independent brake is being used regularly, such as muck-loading or turning on the headlands when cultivating. If this is not done regularly, one brake will wear more than the other and when both are used together the unused brake will lock before the other.

Brake shoe renewal

New brake shoes are usually necessary when either all the adjustment has been taken up, or the shoes are worn to the rivets. In the second case the rivets will score the brake drums if the shoes are not renewed and will impair brake efficiency. It pays to remove the drums occasionally to check wear and blow lining dust out.

To renew the shoes stand the tractor on a hard, level base, block both front wheels and one rear wheel to prevent the tractor rolling off the jack. Release brakes and loosen the adjusters.

Jack up the tractor, remove wheel and place axle stand under the axle as an additional safe-

Fig. 4.83 One way of releasing a tight drum retaining screw is to tap it round with a small chisel.

Fig. 4.84 Gently tap the drum with a soft-faced hammer to remove it. Do not hit the backplate.

Fig. 4.85 Push-off screws.

guard. Clean dirt off the drum and slacken the screws holding the drum. If these are tight use a small chisel to release them (Fig. 4.83). Remove drum. If it is tight tap it gently round the edge with a copper or hide hammer to ease it off. Most drums are cast-steel or iron and may break if hit with a steel hammer (Fig. 4.84). Some drums have threaded holes tapped in them in which push-off screws can be inserted to remove them (Fig. 4.85).

DANGER never inhale or swallow the dust from brake linings.

Make a drawing of the position of the brake shoes holes in which the springs fit on the shoes, and what colour the springs are. Springs are

Fig. 4.86 Lever the spring away with a screw-driver, noting which holes they are in.

colour coded according to strength and must be returned to their original position or if exchanged, replaced by one of similar colour.

Springs can be removed by lifting them from the brake shoe with the tip of a screwdriver (Fig. 4.86). Remove shoes and adjuster. Clean off dirt from backing plate with paraffin and dry off. Apply a zinc-based grease to all bearing surfaces, shoe pivots and adjuster threads. Do not use ordinary grease as it may run on to the linings.

Replace new shoes and springs according to your diagram. If a strong spring is difficult to reposition, double a piece of baler twine round

Fig. 4.87 Strong springs can be pulled into position with the aid of a length of baler twine.

the spring's hook and wrap the string round a spanner.

You can then get extra leverage to extend the spring (Fig. 4.87). Remove the string once the spring is in position.

Ensure the adjuster turns freely if it is the threaded type and lubricate with a smear of the zinc-based grease. Make sure shoes are correctly fitted, so that when the adjuster expands the shoes expand. Clean dust and grease from the brake drum.

If the drum is badly scored by worn rivets replace it. If the scores are not too deep take it to a dealer who has a lathe and he will shim it and remove the scores. Fit the screws into the drum and tighten them up.

Replace the wheel on the axle and the brakes are ready for adjusting.

Disc brakes

Many tractors are now equipped with wet (oil immersed), inboard-mounted disc brakes. These cannot be inspected or changed from the outside; the axle has to be split. Nevertheless, adjustment to take up wear and keep them 'balanced' is necessary. The free travel on the foot pedal should be adjusted to manufacturer's recommendation and each disc linkage should be adjusted so that the tractor 'pulls straight' when test braked. This is usually achieved by shortening the linkage which operates the brakes. Figure 4.88 shows a typical disc brake adjuster.

Fig. 4.88 Typical disc brake adjuster.

PUNCTURES

Many farmers rely on a tyre specialist to repair punctures, yet in most agricultural tyres it is a fairly straightforward job which need not take long.

You will need an air compressor, a selection of patches, suitable rubber glue, a roller to apply the patch, a valve-core extractor, two strong tyre levers and a bead breaker.

Fig. 4.89 Removing the tyre bead from the rim.

The bead breaker is used to remove the bead of the tyre from the wheel rim (Fig. 4.89). It usually consists of a steel bar over which a heavy tube is fitted. The purpose of the tube, which is closed at its top, is to deliver blows to the bar. Should a bead breaker not be available, the bead could be removed with a steel wedge and sledge hammer though this can lead to damage to tyre and wheel.

The bead breaker is the only expensive part of the puncture repairing equipment but it could be made in the farm workshop. Close the top end of the tube by inserting a short length of close fitting bar and weld it in place. The weld must be strong enough to withstand constant hammering, and this is where many home-made breakers fail. The tool will be easier to use if it is made so that the bar and tube cannot separate.

Before looking for the puncture, check that

the valve is not causing the trouble. Reinflate the tyre and watch for bubbles when water or saliva is put on the end of the valve.

Otherwise remove the wheel from the tractor and place an axle stand under the tractor in case the jack should fail when the wheel is off. Remove retaining ring and the valve core to free any air remaining in the tube. Place the wheel on the ground, valve side down, then the wheel will only have to be turned over once.

Force the tyre bead away from the wheel rim with the bead breaker. Turn over the wheel and remove the other bead from the wheel rim. Insert a tyre lever either side of the valve and lever the tyre over the edge of the rim (Fig. 4.90). To make this easier, force the bead opposite the valve into the well of the wheel. Always lift the tyre first where the valve is located.

After the whole tyre has been levered over the rim, it should be supported on blocks so that the tube can be easily removed. Do this carefully because the tube may be attached to the inside of the tyre by a nail and could easily become torn.

Fig. 4.90 Use levers to remove the tyre, starting near the valve.

If a hole is not obvious, replace the valve and reinflate the tube and if necessary the tube must be placed under water to locate the tell-tale bubbles. Mark the hole but do not attempt to repair a tube while it is still inflated because the repair patches are designed to stretch with the tube as it is inflated.

Place the tube on a clean, smooth surface and thoroughly clean the area around the hole with a wire brush (Fig. 4.91). Better still, use a buffer cleaner solution (Fig. 4.92). After cleaning do not touch the tube again with your hands, otherwise the glue may not stick properly.

Apply the glue and allow it to dry (Fig. 4.93). Again, do not touch the glued area or allow dust to settle on it. Usually the glue must be allowed to dry completely. Meanwhile, check the inside of the tyre for damage, in the area around the puncture.

Fig. 4.91 Use a wire brush to roughen the tube.

Fig. 4.92 Clean the tube with a buffer cleaning solution.

Fig. 4.93 Glue can be applied by finger.

Should the tyre be split or cut, it must have a gaiter glued, or better still, welded, inside. A

piece of old inner tube or wellington boot can
be used as a temporary gaiter, but this must be
replaced as soon as possible.

Even a small cut on the inside of a tyre can
pinch the tube as the cut opens and closes when
the tyre rotates under load. Do not gaiter tyres
fitted to road vehicles other than tractor-drawn
trailers, because it is *dangerous and illegal.*

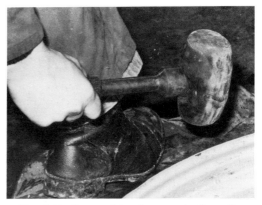

Fig. 4.95 Use a rubber hammer to force the
tyre on to the rim.

be crushed should the tyre suddenly jump on to
the rim.

The tyre should be fully inflated then completely deflated so that the tube can assume a
natural position within the tyre. Finally reinflate
the tyre to the correct pressure and check the
valve for possible leaks.

When inflating a tyre, stand clear in case it
should blow off its rim. Even new tyres have
been known to do this. Tyres retained by locking rings should be inflated in a cage, as the
locking ring can easily fly off during inflation
(see Fig. 4.96).

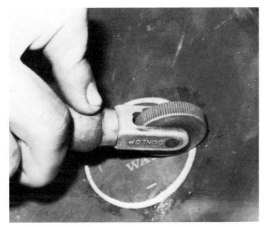

Fig. 4.94 Use a roller to apply the patch.

When the glue is dry, roll on the patch without touching the glued area on the contact side
of the patch (Fig. 4.94). Always use a patch of
adequate size. As a guide, a nail hole requires a
patch of about 25 mm (1 in) in diameter. Patches
of the self-vulcanising type are the easiest to use
and give best results. After applying the patch,
recheck the tube for leaks and then replace it,
so that the valve points towards the valve hole.

It is possible to fit a tube the wrong way
round, and it will soon become damaged or the
valve may disappear back through its hole in
the rim while the tube is being inflated. This
can easily occur when repairing punctures in
tractor front tyres.

Some people advise replacing the valve and
inflating the tube slightly before refitting the
tyre bead to the wheel rim, to prevent it being
trapped between tyre and rim. But when the
tyre is knocked on with a rubber hammer, there
is little chance of the tube being ripped by a
tyre lever (Fig. 4.95). Always start at a point
opposite the valve.

Sometimes, getting both tyre beads on to
their rims again is difficult. So deflate the tube
again and lubricate the beads with a rubber
lubricant or hand soap and try again. *Do not*
use oil and never place fingers between the bead
of a half-inflated tyre and the rim or they will

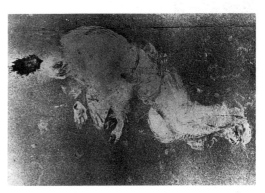

Fig. 4.96 Probably the most unpleasant
photograph of the consequences of a tyre
burst. The picture shows the outline of a
Swedish tyre fitter on the bay ceiling.

It is possible and sometimes preferable to
remove the tube from a wheel with the wheel
still on the tractor. This is a much safer method
than removing the wheel where it is not 100%
safe to use a jack.

Some farm machines are now fitted with
tubeless tyres. One way of repairing these tyres
is to remove the tubeless tyre valve from the trim

with a pair of pliers or failing that, cut it off and then fit a tube of the correct size in the tyre. A split tyre will require a gaiter before the tube is fitted.

GASKETS

Gaskets in engines seal surfaces that will not mate perfectly on their own. When two such surfaces are bolted together, the compressed gasket takes up small irregularities in the surfaces and makes a joint that will keep lubricant and water in and dirt out. Where oil or water are likely to come in contact with the joint, gaskets are made of cork, oil-impregnated paper or rubber. Copper and asbestos, or asbestos alone are used in hot parts of the engine like the exhaust manifold or cylinder head.

Oil or water creeping out of an engine through a badly-fitted gasket may mean an expensive dismantling job. For engine overhauls or cylinder head gasket repairs, your dealer will supply a kit with all the gaskets required for the job. This is often cheaper than buying each gasket separately as required. Before fitting a

Fig. 4.98 It is essential to clean the surfaces down to the bare metal. Use a scraper made out of an old file or steel rule to remove as much as possible of the old gasket and rust . . .

Fig. 4.99 . . . and then finish off with a stiff wire brush to leave a smooth, even surface. If it is badly rust pitted, remove the studs and file down the surfaces until smooth and flat.

Fig. 4.97 If the seal between the two parts is difficult to break, tap the cover gently round the edges with a soft hammer in the direction it comes off until the seal breaks. This plate was the back cover of a cylinder head water jacket and was well rusted on.

new gasket clean the parts to be sealed very carefully (see Figs. 4.97, 4.98 and 4.99).

If no suitable gasket is available use gasket

paper or card, obtainable from your dealer. Ordinary strong brown paper will do. Hold the paper firmly over the surface to be covered and, with a soft hammer, tap it gently round the edges, bolt holes and oil-ways until the shape is cut out. Clean up the edges with a sharp knife or scissors (Fig. 4.100). If the gasket you are making is particularly complex, it may be better

Fig. 4.100 Marking out a gasket.

Fig. 4.102 The hollow punch in operation.

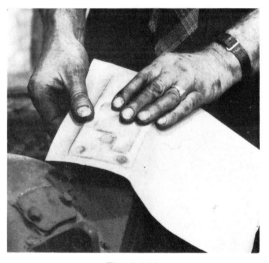

Fig. 4.101

Fig. 4.103 Using a ball-bearing to cut a hole.

to press the outline of the plate out with an oily finger and then cut round the mark with scissors (see Fig. 4.101).

Use a hollow punch, known technically as a wad punch. Figure 4.102 illustrates one being used to cut out a series of neat 8 mm diameter holes.

The gasket should be resting on the end of a smooth wooden block and the punch should be used parallel to the grain of the wood. Never use a wad when the gasket is lying on a hard surface or the punch will be damaged.

Another method is to use a ball-bearing by resting it over the hole, but on top of the paper (see Fig. 4.103). A sharp blow from a hammer will produce a neat hole.

Ensure that the ball-bearing is big enough not to be forced in to the hole. With aluminium components, mark the position of the holes then use an old casting to strike the ball-bearing against. Alternatively drill a suitable piece of steel for this purpose.

When fitting a gasket that needs to be held in position while you hold an engine part, as with sump gaskets, a smear of light grease will stick the gasket in position while you bolt the part.

Use a tapered tool to line up the two parts and the gasket and make sure all the bolts are started in their threads before tightening down.

TRAILER HOSES

Trailer hydraulic hoses are often snapped through drivers forgetting to disconnect them from the tractor or the pipe fouling the tractor linkage. If the break is near the end of the pipe the coupling can be taken off and refitted to the pipe. This is a simple workshop job, but cleanliness is essential as dirt entering the joint could cause leaks in the connections or damage the seals.

Hold the hose in the vice and square up the broken end of the pipe. Use a fine-toothed hacksaw for this job and also to cut the outside rubber coating to the wire braiding, taking care not to damage the wire and weaken the pipe (Fig. 4.104).

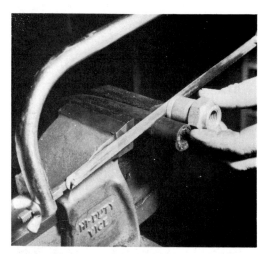

Fig. 4.104

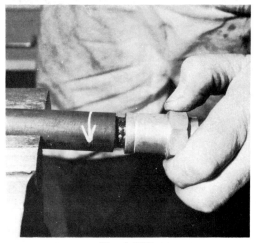

Fig. 4.105

Slit the rubber cover with a sharp knife and peel it off. Next, screw the female end of the coupling on to the pipe (Fig. 4.105), until the hose just bottoms in the fitting. The female half of the coupling has a left-hand thread. Before fitting the insert lightly lubricate the threads and the inside of the hose, to avoid twisting or tearing the hose and to assist assembly. Still holding the pipe in the vice, screw the insert into the female half until the nut on the insert bottoms on the coupling (Fig. 4.106). The insert is tapered and will tighten the hose against the female half as it is screwed on, making an oil-tight seal.

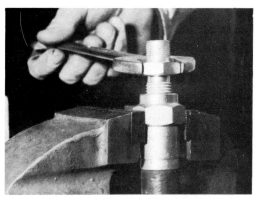

Fig. 4.106

Fig. 4.107

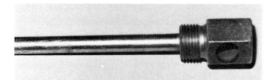

Fig. 4.108 Special mandrel for fitting inserts into sockets.

Screw the tractor coupling on to the insert and tighten up (Fig. 4.107). When refitted to the trailer and operational, inspect for leaks.

Persuading the insert to catch on some hose-pipes can be difficult, unless a special fitting tool is used (Fig. 4.108). This fits inside the insert and extends out through the bottom.

It helps to push out the sides of the hose so that the nipple can enter it more easily. The single wire hose which does not require the outer rubber to be removed always needs the fitting tool.

These tools also screw on to the end of a nipple with a swivel to prevent the nut revolving when it is screwed into the socket. A cheaper solution is to use an old adaptor (Fig. 4.109).

Where a straightforward join has to be made in the middle of a hydraulic pipe, another simple fitting is available (Fig. 4.110). It consists of an externally grooved pipe which pushes into each end of the pipe to be joined (Fig. 4.111). A pair of clamps which are internally grooved are then bolted on to the outside of the hose to prevent the steel pipe from being pulled out of the hose (Fig. 4.112).

Fig. 4.111 Fitting a hose joiner.

Fig. 4.109 Use of an adaptor to fit an insert with a swivel end.

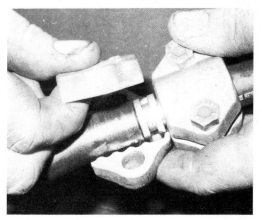

Fig. 4.112 The plates hold the pipe in place.

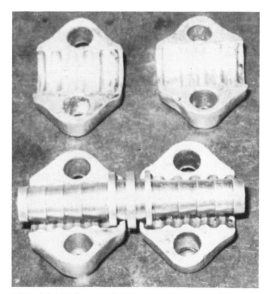

Fig. 4.110 Simple fitting for making a straight-forward join.

HYDRAULIC COUPLING

The most common fault with external hydraulic couplings is leaking valves, especially where slack maintenance has allowed the oil to become dirty. New 'O' rings and valve sleeves can be fitted on the farm, and will save several pounds on replacing the complete coupling.

No special tools are needed. Parts required are a new valve sleeve and 'O' ring for the tractor coupling, and a new 'O' ring and valve for the implement coupling.

To repair a tractor coupling, remove the valve from the fitting bracket by unscrewing the oil pipe and slackening off the locking nut

(Fig. 4.113). Then clean the valve in a suitable solvent, such as paraffin (Fig. 4.114) before removing the spring locking clip securing the valve assembly with a pair of long-nosed pliers. Turn the screw clockwise to release the valve, which is removed from the other end of the coupling. Take out the valve sleeve and spring at the same time and clean thoroughly (Fig. 4.114).

Fig. 4.113 Remove the coupling from the tractor by disconnecting the oil pipe and slackening the lock nut next to the carrying bracket.

Fig. 4.115 The new O-ring is fitted to the valve sleeve and the spring and sleeve replaced in the valve body.

Then fit the new valve sleeve and 'O' ring. The 'O' ring will slip easily into the groove on the valve sleeve if it is smeared with soft soap.

Don't forget that the valve screw has to be turned anti-clockwise when re-fitting. Make

Fig. 4.114 The coupling must be thoroughly cleaned to prevent any trace of dirt getting into the inside of the valve assembly before it is replaced.

Fig. 4.116 The trailer end coupling has an Allen key in the side. This is removed before the coupling is placed in a vice and a pair of circlip pliers used to unscrew the threaded insert from the valve body.

sure the spring locking clip has been replaced before the coupling is fitted back to the tractor.

To work on an implement coupling, take the female coupling from the trailer or implement, and remove the small Allen screw on the side of the rounded part of the body. Then put the coupling in the vice, and use a pair of circlip pliers to unscrew the threaded insert from the valve body. There is a special tool for this job, but circlip pliers are a good second best (Fig. 4.116).

The two halves of the coupling can then be unscrewed by hand to release the valve assembly and pressure spring contained inside. Clean all parts thoroughly.

The 'O' ring can be seen in the bottom of the round section of the coupling. Take it out with a small screwdriver, and fit the new 'O' ring after soaping it. This done, fit the new valve and reassemble the coupling.

In both cases, the entire operation need take no longer than an hour — usually far less — and the result will be a valve coupling that should see plenty more action.

HYDRAULIC CYLINDERS

While a hydraulic cylinder is still on its machine, try to find the cause of a suspected problem. Check that an oil leak is due to faulty seals and not to a cracked end cap (Fig. 4.117) or a leaking ram. Rams which sometimes leak at the pin hole can be cured by welding. Should the piston seals leak on a double-acting cylinder, it will creep or become sluggish.

Before dismantling a cylinder, drain out the oil and plug the oil connections to prevent ingress of dirt. Thoroughly clean the outside of the cylinder. Any grit that gets into the cylinder will not only ruin the new cylinder seals, but will probably find its way into the rest of the hydraulic circuit.

To dismantle the cylinder, first remove the end cap, which may be retained by bolts but usually unscrews. A special tool is usually available, but most caps can be removed by carefully unscrewing with a large pair of stillsons. Never hold the hollow part of a cylinder in a vice because it can easily be distorted. Always grip the end of the cylinder by the pin hole. This section is solid enough to be held in a vice. Prevent the cylinder turning by placing a steel bar through the pinhole.

Another method is to place the end cap in a vice and unscrew the cylinder. Hold cap below jaws (Fig. 4.118) and close vice until the cap cannot rotate because of the lugs but is *not squeezed* by the vice. As many end caps are made of aluminium or cast-iron, never strike them with a hammer.

Fig. 4.118 Unscrewing the end cap by rotating the cylinder.

With the end cap of a single-acting cylinder off, the stop on the end of the ram must be removed. Double-acting cylinders will need the piston removed. Stops and piston are usually screwed on, but may be retained by circlips or roll pins. Again, special tools are recommended for these components, but they can normally be unscrewed, either by holding them between wooden blocks in a vice or with a pair of stillsons, using a cloth to prevent damage by the jaws of the wrench.

Sometimes the holes intended to be used with the correct tool can be employed by inserting two old screws and then gripping the screws in a vice while the ram or rod is unscrewed (Fig. 4.119).

Sometimes it is possible to remove the end cap over the outer end of the rod or ram, or

Fig. 4.117 Cracked end cap.

Fig. 4.119 Makeshift way of holding a piston using two old screws.

pin hole — but this is bad practice because the area around the pin hole will be distorted and will damage the new seals when the end cap is replaced.

The main seals are usually O-rings or a tough lip seal. Here is the procedure for replacing them:

• Lay out the old seals on the bench in the order of assembly, and facing the correct way. They can then be used as a check to ensure that the replacement seals are fitted properly.

• Replace any O-ring under the piston (Fig. 4.120). This prevents oil creeping past the

Fig. 4.121 Seal the piston threads if necessary.

threads of the piston. Often the threads themselves require extra sealing with PTFE tape or a sealing compound, when the piston is replaced (Fig. 4.121).

• Some seals require warming in boiling water before they can be fitted. Check before damaging a tough seal by trying to fit it cold.

• Check that the wiper seal end of the cap is correctly fitted. This seal is to exclude dirt, and

Fig. 4.120 Replace any O-rings fitted under the piston.

Fig. 4.122 Use the old seal to knock the new one into place.

KEYS, KEYWAYS AND TAPER-LOCKS

is normally fitted the opposite way round to an oil seal. Use the old seal to knock the replacement into place (Fig. 4.122).
● Lubricate all the seals before assembling the unit.

When the cap is screwed fully on to the cylinder the oil ports on the cap and cylinder may not be in the correct position in relation to each other. This can be solved by placing shims of varying thickness in the end cap (Fig. 4.123) so that it will tighten down in a different position. This can be a problem when a replacement end cap has been fitted. A new end cap should be screwed on to the cylinder and the necessary shims fitted *before* the seals are fitted. When the cap is finally fitted, a little sealing compound on its threads may help to prevent leaks.

Fig. 4.123 Fit shims under the end cap if oil ports do not line up.

KEYS, KEYWAYS AND TAPER-LOCKS

One of the most common ways of joining a gear or sprocket to a shaft so that they revolve together is by means of a key. Temporary fastenings, the keys are always made of steel, as they are subjected to considerable crushing and shearing stress as the shaft turns.

The key is always made of softer steel than the shaft, being cheaper to replace.

A keyway consists of a recess in a shaft or hub to accommodate the key. There are five main kinds of sunken key — square and rectangular taper, square and rectangular parallel, and woodruff.

Taper keys have a standard taper of 1:100. They are measured at the larger end for thickness, are uniform in width and have either square or rounded ends. They are fitted by driving the

key into the keyway until the taper tightens against the component and prevents the part from moving.

Taper keys can be difficult to remove. The best method is to knock the sprocket down the shaft, take out the key and pull the sprocket off. This can be difficult if a rust seal has formed between the key and component, or if there is insufficient room to allow the sprocket to be knocked down the shaft. If this happens try heating the sprocket boss with an oxyacetylene torch to expand it and pull out the key with pliers or grips.

To prevent these problems arising use a gib-head taper key. This type has a raised head to allow it to be knocked in and prised out without damaging the taper. To extract the gib-head key, make a taper drift out of 6 mm thick steel plate about 150 mm long, 50 mm wide at one end and tapering to 6 mm at the other. Place the narrow end between the key head and component and hit the wider end. The taper on the drift will force out the key without

Fig. 4.124 Gib-head keys should be extracted with this home-made tapered drift.

Fig. 4.125 Sharp taps with a hammer soon remove stubborn keys.

damaging the key or component (see Figs. 4.124 and 4.125).

Parallel keys are virtually the same as taper keys except that, as their name implies, they are parallel throughout their length. They are used mainly where a pulley or gear has to slide sideways on a shaft until the belt or chain that fits on it is in line. Then the component can be washered up and held in position by a split-pin.

Main snag with these keys is that they tend to 'fret' or move in the keyway and try to turn over, particularly if the drive is often reversed. This results in wear on both the key and keyway edges, and the key becomes a loose fit in its groove and eventually shears (Fig. 4.126).

Woodruff keys are deeply sunk and are easily adjustable to take up any taper in the hub of the component. They are shaped as part of a circle and the keyway in the shaft is cut on a milling machine to give a recess of the same diameter as the key.

Advantages are that the deep fitting prevents the key turning over, and it will tilt slightly to accommodate any small taper in the boss to which it is fitted. But because of the much deeper keyway the shaft can be weakened at that point.

Other types of key are the feather, which can lock a unit anywhere on a splined shaft, and flat, hollow and round keys, used for light work (Fig. 4.127).

Fitting a key

The secret of fitting a key properly is to ensure that the correct sized key is used for the job and that it fits well in its keyway. A sloppy fit will wear the shoulders on the recess in the shaft or gear hub, and the key will try to run over.

First make sure that the hub fits the shaft. Remove high spots and rust on the shaft with emery cloth and try the hub on the shaft. Next clean up rough edges on the key and shoulders of the keyways with small, fine file and try the key for size.

If the key moves sideways more than about 0.1 mm, a larger key is needed. If the shoulders of the recess are badly worn, a new shaft is

Fig. 4.126 On the right is a loosely fitted key which began to move in the keyway and became worn.

Fig. 4.127 Several types of key — the gib-head, woodruff, parallel and feather, looking from the top.

Fig. 4.128 Use a flat-ended punch to tap the key into its recess.

necessary or the keyway can be recut on another part of the shaft.

Tap the key squarely into its recess (Fig. 4.128). The same amount of the key should be showing above the shaft for the whole of its length. Finally slide the sprocket on the shaft, line the keyway up with the shaft, and tap it over the key until it is in its correct position (Fig. 4.129).

Fig. 4.130 Typical taper-lock bush assembly.

Fig. 4.129 Knock the sprocket on to the shaft and over the key with a soft-faced hammer.

Mr Brian Parker, a welding contractor, has evolved a simple method of renovating worn keyways in shafts.

All that is required, he says, is a dummy key made of copper and an electric welder. The copper key needs to be the shape and thickness of a new key.

Place the dummy key tightly in the keyway and build up the shaft on either side of the key with weld. The copper will not fuse with the weld metal and will prevent the keyway being filled.

When the shaft has been built up sufficiently remove the key and file off the excess metal to give a neat repair.

Special keys

Keys are often made deliberately soft so that in the event of a shock load in the drive, the key will shear. Such a key must be replaced with one made from the same material.

The taper-lock bush is an ingenious method of fixing a hub to a shaft (Fig. 4.130). It has a parallel bore in which a key-way is machined. Because the bush is split, the exact diameter

of the bore can be changed. The bush becomes tight on the shaft when it is pulled into the hub which is tapered to match the external taper of the bush. The bush is held in the hub by the screws which have threaded holes half in the hub and half in the bush.

Removal of a hub retained in this way is simple. Remove both the screws which lock the bush to the hub and then use one to push the bush out of the hub by inserting it in the third hole (Fig. 4.131).

Fig. 4.131 Removing a taper-lock bush by inserting a screw in the third hole.

When the bore and keyway of a large expensive sprocket wears badly, one answer is to have it bored out to take a standard taper bush or have a taper lock hub welded on to the sprocket. These are available in many sizes.

UNIVERSAL JOINTS

Gradual wear, accelerated by mis-alignment, will cause the universal joints used in pto drive shafts to become slack. If this is not attended to the bearings will break up and the coupling itself may be badly damaged.

The renewal of bearings and cross piece is not difficult and can be quickly carried out. The new parts required are a cross piece, containing the grease nipple, which supports the two halves of the coupling, and four new cups containing needle bearings, one for each knuckle of the joint. When these parts have been fitted the coupling will be serviceable for a further period of use and costly replacement of the entire coupling will have been postponed.

The job of replacing the worn parts in a universal joint is illustrated in Figs 4.132 to 4.137. Before starting work, mark the yokes so that they can be replaced in their original positions.

Place the universal coupling in a vice and remove the spring clips from each side of the knuckles on the universal joint. Cups are pushed partly out of the yokes by placing one yoke across the vice and tapping the other yokes with a hammer. The large yoke is more cumbersome and should be removed first, before the cross piece is similarly placed across the vice and the cups driven in turn from the smaller yoke (Figs 4.132 and 4.133).

Fig. 4.132

Fig. 4.133

When each cup has been exposed it is gripped in the vice and removed by twisting and pulling the coupling. Next, clean off both yokes and drive one of the old cups through each hole to remove dirt and burrs (Fig. 4.134).

Fig. 4.134

To reassemble, lay out the replacement parts on a clean surface and grease the inside of the cups to retain the needle bearings. Push a new cup into one of the holes on the smaller yoke, but do not push it fully home.

Place the cross piece into this cup from the inside then place the yoke in a vice to squeeze the cup into the yoke (Fig. 4.135).

Fig. 4.135 With the cup in position, place the coupling in the vice and tighten to press the cup into the coupling jaw until it is a flush fit.

A spacer is then inserted between the cup and the vice jaw, which is tightened again to press the cup further into the yoke. This pushes the cross piece well into the opposite hole to facilitate fitting the next cup. When this is pressed home in the vice the cross piece is centred (Fig. 4.136).

Fig. 4.136 Use a spacer to push the cup further into the yoke.

The grease nipple is then fitted and the spring clips replaced. Care should be taken to fit the cross piece so that the grease nipple faces outwards when the assembly is complete. Never over-grease the universal joints or the seal will be damaged (Fig. 4.137).

Fig. 4.137 Finally replace the spring clips and the grease nipple.

VEE-BELTS

Vee-belts consist of woven nylon cords encased in a synthetic rubber compound. Their shape is a truncated vee with an included angle of about $40°$ between the sides.

The pulleys they drive have a groove of the same shape as the belt. As the belt wears it slips deeper into the pulley to continue driving. The belt should be replaced long before it can touch the bottom.

Correct tension is the vital factor with any type of belt. Too much slack will not allow full power to be transmitted and slip will occur. If the pulley is made of alloy, excess slip will cause heating and grooves will form in the pulley sides. These will soon ruin a new belt. Too tight a belt will cause wear on pulley and motor bearings and fractures in the belt casing and cords.

Movement per 30 cm of distance between the pulleys should be about 12 mm. A properly tensioned belt will have a springy feeling. After initial stretching a belt will not need retensioning for a long time. Do not attempt to overtension a belt in advance to overcome this initial stretching.

Machines such as forage harvesters, mills and mixers often have banks of 2, 3 or 4 vee-belts. These consist of matched belts of the same length and if one of them fails the set must be replaced with another matched set. As the belts wear they stretch slightly and replacing one will mean that the new belt will do most of the driving.

When replacing a belt undo the tension adjusters fully to allow the belt to fit over the pulleys. Do not use a screwdriver or tyre lever to get the belt in position, this will weaken the cords and cause fractures in the casing.

If there is no way of telling what length a broken belt was, choose a piece of rope that will sit neatly in the pulley vee, and wrap it round all the pulleys to measure the length. It will give some indication of the length for your dealer. When a new machine is delivered to the farm make a note of numbers printed on the belt. A useful tip is to paint the number on the inside of the guard. When you put a machine away at the end of the season, release the tension on all the belts. This will reduce the amount of 'set' and rough running when you start the machine next time.

A few degrees of misalignment in a belt pulley will shorten belt life. It will cause uneven wear, the belt may roll over in the pulley groove and power transmission will be reduced. Or it may throw all the load on to one side of the belt, causing the cords to stretch on that side.

Use a straight edge to check alignment. Spin the pulleys by hand to see that they are not bent and check for chipped edges.

When the straight edge is laid across both pulleys it should touch all four edges. Adjust the alignment until it does.

With many machines altering the tension

also alters alignment, with holes slotted one way to take up fore and aft movement for tension, and the other way for sideways movement for alignment.

Maintenance of belts consists of keeping them free from oil and grease, otherwise they perish, swell and rot. Ensure that the pulleys are in good order. Store short belts by hanging from a nail in a dry corner of the workshop out of sunlight. Long belts can be folded and stored flat on a shelf.

Inspecting Vee-belts properly may involve removing the belt and turning it inside out. The crack in the belt shown in Fig. 4.138 could not be seen whilst it was still fitted.

or tear away from the belt. As the oil softens the rubber, the belt will stretch more.

Weather: Sunlight, as well as heat will shorten the life of the belt. Whenever possible, store machines so that a minimum amount of light reaches their belts.

Do not hang heavy belts from a single hook, as this causes distortion. Use two or three hooks. Long belts are best folded for storage.

Keep matched sets together.

Belts stored on a machine should be loosened to reduce the risk of them developing a permanent set. Occasionally manufacturers recom-

Fig. 4.138 Turn a belt inside out to inspect it properly.

Causes of damage

Fitting: Many belts are damaged by being stretched or even cut during fitting. Never use a pry-bar or tyre lever to force a belt over the edge of a pulley. Where a belt is still tight to fit after slackening the tensioning mechanism, lay the belt on one pulley and over the right-hand edge of the driving pulley. Rotate the pulleys clockwise and the belt will jump into place.

Slipping: This is usually due to lack of tension, but can be caused by worn pulleys or overloading. Severe slipping is usually accompanied by belt squeal, as the belt takes up the drive.

Heat: All belts tend to heat up during work due to the energy absorbed as they stretch. Unless the belt is overloaded, this heat is unlikely to affect it. Too much heat generated due to slippage or lack of ventilation can cause the belt to stretch and crack. Such cracks usually first appear across the bottom of the belt.

Oil: This causes the rubber to swell and become distended, and pieces of rubber start to break

Fig. 4.139 Hold the belt firmly on the ground and twist it inside out.

Fig. 4.140 Keep twisting for one complete turn.

mend that belts be left on the machine fully tensioned.

Folding a vee-belt

An item which seems to come 'alive' when it is being folded is the vee-belt — particularly one with a long, large section. As usual, there is a right and a wrong way (see Figs 4.139 to 4.142).

Fig. 4.141 Pull up until the two ends come together.

Fig. 4.142 Fold the ends over each other and the belt will then concertina into a compact coil.

Fig. 4.143 The principal chain types: hook-link (top), roller chain (centre) and precision roller chain.

CHAINS

Chain drive, common in agriculture, is used mainly where a positive drive is needed (Fig. 4.143).

Three main types of chain are in common use. The hook-link type, made of malleable cast-iron or pressed steel, is generally used on low-speed work with rough cast sprockets (Fig. 4.144).

Fig. 4.144 A heavy hook-link chain showing the proper direction of travel.

The roller chain, for higher speed drives, is made of steel treated to resist corrosion and can replace malleable and pressed steel chains. It will run on the same sprockets as malleable chains and in some cases steel chains but a trial fit on the sprocket must be made first.

Lastly there is the precision roller chain, used in high speed applications. It will take higher loads than the other types. An accurately machined sprocket is essential with this chain.

Because of their reliability and simplicity chains are often neglected and cause inconvenient breakdowns. Maintenance is easy with regular inspections to check links and rollers. Lubrication requirements depend on the operating condition. If the chain runs in an oil bath, topping up is all that is required apart from recommended oil changes.

Repairing a hook type chain is easy — simply remove the damaged link by bending the chain and sliding one link from the other. A new link can then be fitted.

Repairing a roller chain is more difficult because of the rivets that join one link to

another. It is possible to use a chisel and punch to remove the pin, but this often results in damage to the other links which causes more trouble.

A quicker and easier way is to use a bearing pin extractor (Fig. 4.145).

Fig. 4.145

After clamping it on to the roller of the pin to be removed, screw the tommy-bar down and push out the pin with the centre punch. A new link can then be fitted.

The same procedure can be used to lengthen or shorten a chain, but, instead of fitting only a connecting link, a cranked link is fitted as well to give the correct length.

New links fitted to old chains

A common mistake made when dealing with chains is to use a new link or section of links to repair a stretched chain. This comment also applies to the use of a new joining link to join an old worn chain. If this is done the chain will be subjected to a shock load every time the new link or links engage the sprocket. It is surprising how quickly this shock load will cause the whole chain to deteriorate therefore it is unfortunate that sometimes a new link is the only one available in the farm workshop.

If a roller chain is operating in dusty conditions, as on a combine, then a graphite grease bath is needed.

First clean off all dirt with paraffin and a stiff brush, moving the joints to ensure that caked dirt inside is shifted.

Heat up some graphite grease in a shallow tray until the grease melts. Immerse the chains in the liquid grease for 10 minutes moving the chains regularly to allow the grease to penetrate all the rollers and links. Then hang up the chain and let it dry off. This will be sufficient lubrication. Never spread oil over a roller chain in dusty operating conditions; the dust and the oil will form a grinding paste that will rapidly wear the chain out.

For malleable and pressed chains and those in non-dusty conditions simply soak the chain in oil after cleaning and let it drain before refitting.

Wear in malleable or pressed steel chains is easily shown by broken or cracked links and sprockets which must be replaced.

Check roller chain wear by measurement. If over 25 links the chain has worn more than half a link's length, fit a new chain. Check the sprocket for wear — fitting a new roller chain to a worn sprocket is a waste of money. The misshapen teeth will quickly wear the new

chain. If the teeth show signs of hooking then renew the sprocket as well.

Tension all types of chain so that there is about 12 mm of movement on the slack side with shaft centres about 300 mm apart. Avoid over-tightening chains, or bearings, chain and sprockets will wear excessively.

When fitting chains to the sprockets make sure the connecting link or split pin is fitted the right way as shown illustrated, otherwise the chain may split and fall off. Also make sure that safety guards are replaced after chain adjustments (Figs 4.146 and 4.147).

PULLERS

Few machines can be dismantled completely without the use of a puller. This tool can be made in the workshop to do a particular job as well as any you can buy. Alternatively, a standard tool can be adapted to perform several jobs. Pullers are obtainable in many different shapes and sizes. When made of strong steel they all have one thing in common — they are expensive. The following are some uses for pullers with, where possible, an alternative method of tackling the job.

Fig. 4.146 On a roller chain, align the split pins to point forward.

Fig. 4.148 Removing an engine front pulley.

Fig. 4.147 The corresponding alignment for links on a precision roller chain.

Pulleys

Many pulleys are damaged during attempts to remove them with bars or tyre levers. Fig. 4.148 shows a tractor crankshaft pulley being pulled off by inserting two screws in the holes already tapped in the pulley. This illustrates two points. First, the puller can be fitted with two or three legs. Second, the forcing screw of the puller has been allowed to push against the inside of the crankshaft because this pulley is not a tight fit.

When removing tight-fitting pulleys, a shaft protector must be used to prevent damage to the threads in the crankshaft. Third, pulleys can be removed without a puller by inserting long screws in the tapped holes that protrude far enough from the back of the pulley to push it off the shaft. Be sure to place steel plates for the screws to push against, or they may damage the casing.

Fig. 4.149 Pulling a gear off its shaft.

Fig. 4.150 Note that the shaft has a hole to accept the nipple of the puller.

Gears

Figure 4.149 shows a gear being pulled off its shaft. This is a straightforward job providing the following points are observed: check that the shaft has a small hole in the end to accept the nipple on the puller's forcing screw (Fig.

Fig. 4.151 Typical hydraulic puller.

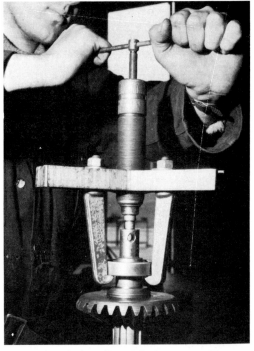

Fig. 4.152 A puller with three sets of legs.

4.150). Otherwise drill one or else a short length of steel shaft drilled to take the nipple on the puller must be inserted between the puller and the shaft. Failure to do this will result in the flattening of the nipple on the end of the forcing screw. The purpose of these nipples is to help keep the puller centralised.

Should the gear be tight to remove, strike the end of the forcing screw while the pressure is on (Fig. 4.149). Keep tightening the screw as the end is struck, but *never* strike the screw of a hydraulic puller (Figs 4.151 and 4.152). Hydraulic pullers can be recognised by the small screw in the main body which usually exerts much more pull than a conventional puller which uses a forcing screw. Should a hydraulic puller fail to remove a stubborn component, try striking the component as near as possible to its centre. When the component is held on a taper, try striking the taper while the puller is under load.

Bearings

Often when bearings have to be removed from a shaft without damage, a set of bearing puller plates must be used (Fig. 4.153). These are a worthwhile investment because they can be used as shown or a conventional two-legged puller can be used to pull on the outside of the plates. Fairly complicated jobs can be tackled. Figure 4.154 shows a Ford axle bearing being removed from its shaft. The two long pulling bolts have been made to suit this particular job.

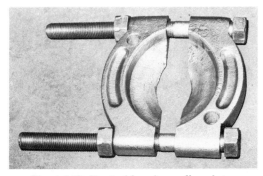

Fig. 4.153 Typical bearing puller plates.

Sometimes the outer races of bearings can be difficult to remove, especially when located where there is nothing for the forcing screw of a puller to push against. Here (Fig. 4.155) the forcing screw of a puller has been removed and the shaft of a slide hammer screwed into the puller bridge piece.

When the weight on the slide hammer is jerked upwards against the stop on the end of the shaft, the pulling action should remove the bearing race. This puller has legs attached to the bridge piece near the centre so that a threaded bar can be fitted between the upper ends of the legs to pull them together so that the lower ends of the legs will always tightly grip the bearing race.

A slide hammer can be made easily on the farm and is a useful form of pulling device because almost any conventional puller can be screwed on to its shaft.

A bearing should slide easily on and off the

Fig. 4.154 Removing an axle bearing.

Fig. 4.155 Puller being used with a slide hammer.

shaft and in and out of its housing. This seldom happens as burrs, rust and dirt help to make removal difficult.

If a bearing will not come off easily because of spreading on the end of the shaft use a file and emery cloth to remove the excess metal. If no pullers are available, two tyre levers placed behind and on opposite sides of the bearing may help. Give the levers sharp taps with a hammer in turn to try to move it.

Remember, the pullers or tyre levers must be positioned behind the inner race of the bearing. Clean up any damage and grease the shaft well before fitting a new bearing.

If filing, levers or pullers will not budge the bearing, gas-cut it from the shaft. The problem is to cut the bearing without damaging the shaft. First cut off the outer race, taking care not to burn any plastic, wood, or rubber in the area of the bearing. (Fig. 4.156).

Fig. 4.156 Gas-cut a bearing from a shaft if all other methods fail.

The second cut must be across the width of the inner race. Aim to cut through half the thickness of the race, so that the flame does not touch the shaft.

Use a chisel and hammer on the half-cut to split the race and it will slide off the shaft (Fig. 4.157).

Fig. 4.157 After making the second cut, use a chisel to split the race.

MILKING MACHINES

Less than half the milking machines in Britain operate efficiently. This is mainly due to poor routine maintenance. Some of the faults which milking machines develop remain unrepaired for many months or even years. Often these faults can cause udder infections by prolonging the milking time or producing undesirable pulsation characteristics. Many of the problems which develop with milking machines will be eliminated if the following checks are regularly made. (See Figs 4.158 to 4.173).

Fig. 4.158 Check pulsation rate.

Fig. 4.159 The vacuum level must be correct.

Fig. 4.160 Examine the vacuum regulator.

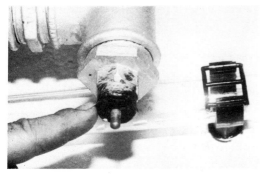

Fig. 4.161 Checking for leaking drain cocks.

Check that the pulsation rate is the same for all units and that it falls within the limits stated by the manufacturers, (this is normally between 50 and 60 pulses per minute). Check the rate by placing a thumb in one cup and counting the number of pulses in one minute. At the same time the condition of the liners should be checked (Fig. 4.158).

Fig. 4.162 Replace perished rubbers on the units.

The vacuum level must be correct. If there is any doubt about the accuracy of the gauge, check it against one which is known to read properly (Fig. 4.159).

The vacuum regulator should be examined and cleaned if necessary. Pay extra attention to the valve setting. The regulator should not start to allow air into the system until the vacuum has reached the correct level. Air leaking in is often indicated by a hissing sound from the regulator (Fig. 4.160).

Air can leak into the system through leaking drain cocks, due to worn valve seatings or washers (Fig. 4.161).

Perished rubbers on the units should be replaced before they give trouble (Fig. 4.162).

Regularly empty the sanitary trap, check its seals and the action of the float valve because if the trap became flooded and the float valve failed, liquid would be drawn into the pump and cause damage (Fig. 4.163).

Fig. 4.163 Regularly empty and check the sanitary trap.

One place air should be allowed to enter the system is through the air bleed in the claw. A blocked air bleed will restrict milk flow from the claw (Fig. 4.164).

The non-return valve in the milk pump should be checked for wear and replaced if necessary (Fig. 4.165).

Should the milk pump start to give trouble, make sure that part of the pipe work has not slipped. It should normally be arranged so that the milk can run down into the pump by gravity (Fig. 4.166).

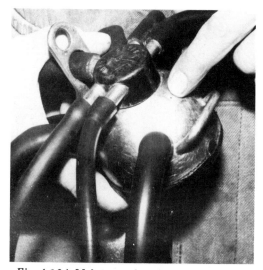

Fig. 4.164 Make sure that the air bleed in the claw is not blocked.

Fig. 4.165 Examine the non-return valve in the milk pump for wear.

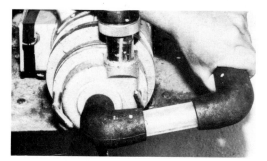

Fig. 4.166 Look for pipe work that may have slipped if the milk pump gives trouble.

Fig. 4.167 Occasionally lubricate the suspension slide of the jar.

Fig. 4.168 Checking the fine adjustment of the milk pump switch.

Electric milk pumps are usually switched on by the milk jar moving downwards and operating a switch. Make sure that jar is free to move by occasionally lubricating the suspension slide (Fig. 4.167).

The fine adjustment of the milk pump switch should not require much attention, but they have been known to slip out of adjustment (Fig. 4.168).

The pulsator filter is easy to forget. It must be changed at the stated interval (Fig. 4.169).

Cluster removers are normally air/vacuum operated. If one shows signs of sticking, dismantle it and thoroughly clean both the piston and the cylinder. Pay extra attention to the piston seal (Fig. 4.170).

Fig. 4.171 Top up the oil in the vacuum pump weekly.

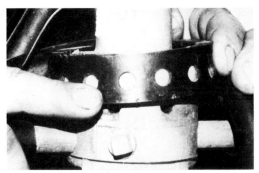

Fig. 4.169 Change the pulsator filter at the stated interval.

Fig. 4.172 Check the tension and condition of the pump drive belt.

Fig. 4.170 Sticking cluster removers must be dismantled and cleaned

The level of oil in the vacuum pump should be topped up at least weekly. Use the recommended lubricant for this purpose (Fig. 4.171).

At the same time check the tension and note the condition of the pump drive belt. If necessary, order a new belt (or set of belts) now, do not wait until it starts to slip badly or breaks (Fig. 4.172).

Fig. 4.173 The vacuum pump should be installed in a purpose-built house.

The vacuum pump should preferably be installed in a purpose built, well guarded, but easily accessible house (Fig. 4.173). Note that its exhaust should always point downwards so that condensation can not run back into the pump when it is stationary.

REASONS FOR BEARING FAILURE

The condition of a bearing can be assessed by listening to it at work. Place a screwdriver on the housing and put an ear to the handle. The various sounds and their implications are shown in table 4.1.

A bearing that can be heard without the aid of a screwdriver is already in poor condition. A noisy bearing in an oil bath needs immediate attention because fragments from it can contaminate the oil and damage the other bearings. Other signs that all is not well are:

A sudden rise in temperature. This could be caused by excessive lubrication or incorrect adjustment. A bearing which should have some axial movement may have been incorrectly adjusted. When the bearing heats up slightly it expands and develops a self-imposed pre-load which causes it to heat up more.

Loss of lubricant. This may be caused by a faulty seal or the bearing heating up the lubricant and forcing it out of the seal.

Failed bearings. Inspection of the bearing's working conditions may reveal the cause of failure. For instance, the rotor bearing of a flail

mower will fail when the machine is operated with one flail missing. This puts an enormous load on the bearings. Tractor front-wheel bearings do not last long when the hub cap is missing or punctured. When removing a bearing, mark all the mounting housings so that they can be re-fitted in their original places. Remove any damaged bearing carefully and clean it with paraffin for inspection.

Rust indicates that the bearing's housing-seal has allowed moisture in and, possibly, that the lubricant has not succeeded in keeping moisture out. This is often seen in neglected front-wheel bearings of tractors used for yard scraping. Bearings that have turned black or dark blue or brown have been overheated.

Cracks are caused when the bearing is fitted in a distorted housing, forced into a housing which is too small or forced on to an oversized shaft. Using a hammer (Fig. 4.174) is likely to

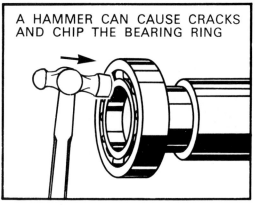

Fig. 4.174 Don't do this.

Table 4.1 Sounds of failure

Sound	Possible cause
Continuous soft purring noise	Normal bearing
Squeaking or metalic sound	Probably lack of lubricant
Metallic tone	Probably the bearing has too little internal clearance. This may have been caused by forcing the bearing on to an over-sized shaft
Ringing clear tone	Indentation in the outer race which may have been caused by incorrect fitting
Regular knock in time with the speed of rotation of the bearing	A damaged race — inner or outer depending upon which one is rotating
Rumbling	A fluted track which may have been caused by the passage of an electric current
Intermittent knock or rattle	Damaged ball or roller
Crunching sound	Dirt in the bearing

fracture the bearing race. But the damage may not show until the bearing has been in use.

Flaking describes the condition when parts of the bearing surface start to lift away from the ball or roller tracks. The main causes are:
• A distorted outer ring caused by a misshapen housing or dirt between the outer ring and the housing.
• Incorrect mounting technique such as applying force to the outer ring in order to put the inner ring on a shaft (Fig. 4.175).

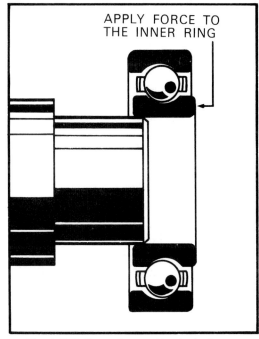

Fig. 4.175 Correct mounting technique — applying force to inner ring.

• Lack of lubrication.
• Dirt. Even soft materials like wood will damage a bearing. Wood which shows any sign of splintering should not be used to tap bearings into place.
• Electric current allowed to pass through a bearing will cause a 'wash-board' on the inner or outer race bearing surfaces. Half a volt is enough to damage a bearing. Do not allow welding current to pass through a bearing. When welding a shaft on a machine, the earth lead of the welder should be clamped to the shaft and not to the most convenient part of the machine.

As well as examining the bearing, check the housing for high spots that would distort the outer ring. Never deliberately damage the outer ring or the housing to secure the outer ring.

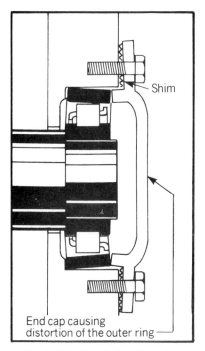

Fig. 4.176 Check for distortion.

When the bearing is retained by an end-plate (Fig. 4.176) check that this is not distorting the outer ring when tightened. This can be prevented by fitting carefully sized shims under the end-plate. The shim thickness is critical. When in doubt, seek professional advice. When

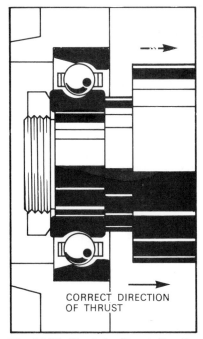

Fig. 4.177 Check for thrust direction.

a bearing has been changed, the thickness of the shims must also be changed. Some bearings are designed to withstand end thrust in one direction only. Check that such a bearing was mounted the right way round — that the thrust is applied towards the more open side of the bearing (Fig. 4.177).

Excessive lubrication can cause a bearing to fail. The lubricant slows down the rolling elements at certain points in their travel and they slide over the bearing surfaces, which causes excessive wear. Too much grease in a bearing can cause it to overheat.

FITTING BEARINGS

Bearing replacement requires careful handling and a degree of mechanical expertise.

Never remove a new bearing from its box except for replacement or maintenance purposes. Study the assembly procedure for the machine so that any seals — often fitted before the bearing — are not left out, and the correct thickness of any shims can be calculated.

The work area should be dust free, and the housing and shaft clean and free from rust. Refitted bearings should have been thoroughly washed in paraffin or white spirit and dried with a non-fluffy cloth. When compressed air is used to dry a bearing, do not allow the air to spin it rapidly as this causes damage. Never wash out a sealed bearing. When the seal is damaged the bearing should be replaced.

Inspection

After cleaning, bearings can be inspected and the condition of the internal surfaces judged by holding the bearing by the inner race and slowly rotating the outer race (Fig. 4.178). If there is any sign of roughness, re-wash and test again. Thrust and taper bearings can be tested by putting the bearing on a clean, flat surface and

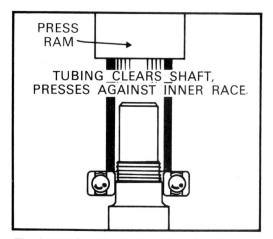

Fig. 4.179 A method of pushing a bearing on to a shaft using a piece of tube to press on the inner race.

rotating slowly while pushing downwards with the hand (Fig. 4.178).

The golden rule is always to press the tight-fitting part of the bearing. Never allow the rolling elements in a race to transmit the force used when fitting. Figure 4.179 shows one method of pushing a bearing on to a shaft using a piece of tube to press on the *inner* race. Figure 4.180 illustrates a larger piece of tube used to press the outer race into a housing.

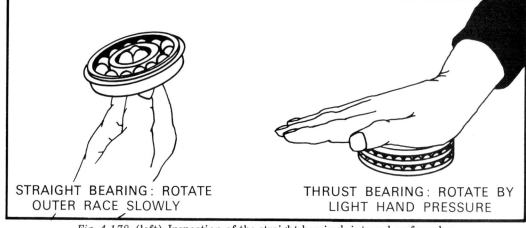

STRAIGHT BEARING: ROTATE OUTER RACE SLOWLY

THRUST BEARING: ROTATE BY LIGHT HAND PRESSURE

Fig. 4.178 (left) Inspection of the straight bearing's internal surfaces by holding the bearing by the inner race and slowly rotating the outer race. (Right): inspection of a thrust bearing.

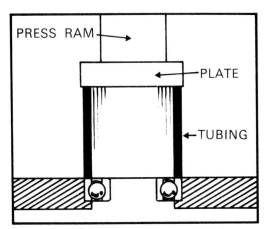

Fig. 4.180 Pressing the outer race into a housing using a larger piece of tube.

Cracks

Never drive the bearings into place with a hammer and punch (Fig. 4.181). Misplaced blows and shock loads on the rings may produce cracks which will not be visible until the bearing is in use.

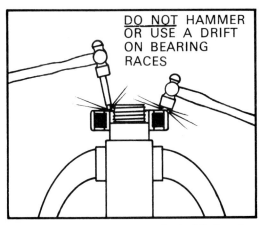

Fig. 4.181 An incorrect method of fitting the bearings into place. Misplaced blows may produce cracks.

Bearings are easier to fit on to a shaft when they are warmed first. Any source of heat except a naked flame may be used.

The safest way is to warm the bearing in an oil bath, but the oil must be clean and have an ignition point of more than 250°C. The vapour is more likely to catch fire than the oil itself, so avoid naked flames.

A bearing allowed to rest on the bottom of the oil bath may become distorted. Always place the bearing in cold oil and allow it to warm up with the oil. Ensure that the bearing temperature does not exceed 100°C, usually sufficient to allow expansion and allow it to slide on to the shaft with little effort.

Another way of ensuring an easy fit is to cool the shaft in a deep freeze before fitting. This procedure may also be used to fit a bearing into a housing by warming the housing and, when necessary, cooling the bearing. An electric light bulb is often good enough when left near the housing for about one hour before fitting. A gas welding set may generate too much heat and crack it.

Keep the bearing pressed hard against the shoulder of the shaft or housing until it has cooled enough to ensure that the bearing will not move as it cools further.

A puller can be adapted to get the bearing on to a shaft (Figure 4.182). Its legs should be fitted with a special plate so that the force is applied to the *inner* rather than the outer race.

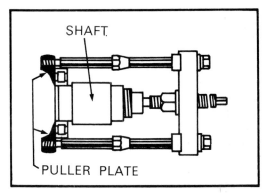

Fig. 4.182 A puller adapted to get the bearing on to a shaft. Force is applied to the inner rather than the outer race.

As a rule, ball or roller bearing housings should be packed about half-full with grease. Never fill the housing completely as this may cause over-heating due to excessive churning and loss of lubricating properties. The grease may become thin enough to leak past the seal.

The bearing should be packed with grease to the outside of the rings, and rotated a few times to displace any excess grease from the rolling elements.

Common mistakes

(1) Pressing the wrong part of the bearing. The rolling elements can easily be damaged, but this may not be apparent until later.

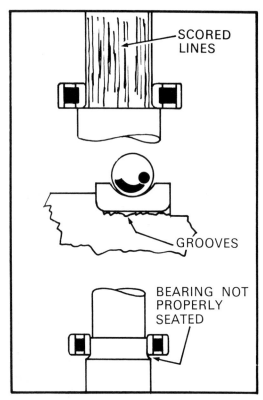

Fig. 4.183 A common mistake — bearings being fitted to an irregular surface such as a scored shaft, may distort one of the races.

(2) Fitting the bearing on to an irregular surface, such as a scored shaft or damaged housing (see Fig. 4.183). Surfaces with high spots on them may distort one of the races, causing early bearing failure.

(3) Damaging the bearing with direct blows from a hammer.

(4) Forcing the bearing on to an over-sized shaft, or into an undersized housing may distort or crack a race.

(5) Over-tightening taper roller bearings causes early failure by pre-loading the bearing excessively. In general, wheel bearings should be tightened until there is slight resistance to the rotation of the wheel, then the adjusting nut slackened until the wheel rotates freely.

(6) Allowing dirt to enter the bearings during fitting.

(7) Interchanging the components of separable bearings. Bearings which come to pieces such as taper roller bearings, must not have their components mixed up. It is also a false economy to try and save time by renewing only the part of a bearing which is easily removed. The new

part will soon be damaged by the non-matching component.

(8) Distorting the bearing by incorrect housing assembly. This can happen when the caps of a split-bearing housing are reversed or bolted together too tightly. Bearings held by an end cap will be distorted when it presses too hard against the bearing. This can be avoided by carefully shimming the end cap.

Often, all the information required to buy a replacement bearing is the maker's name and the number stamped on the side of the rings.

Should the bearing numbers not be legible, most bearing manufacturers will usually be able to supply a replacement when provided with the following information: the type of machine; the position of the bearing on it and its job; all the bearing dimensions (Fig. 4.184) whether metric or imperial; the type of rolling elements — roller or ball; whether it is sealed or semi-enclosed on one side; and whether it is designed to take end thrust.

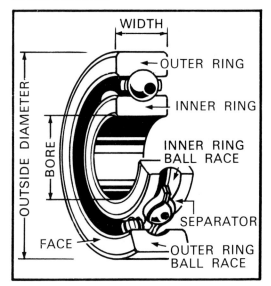

Fig. 4.184 Bearing dimensions.

ADHESIVES AND SEALANTS

Many modern sealants and adhesives used in manufacturing can also be useful in the farm workshop. Most sealants fall into one of four categories:

● *Hard setting types.* These, such as Green Hermetite, are very strong and excellent on threaded joints. They can withstand high pressure but should not be used where any movement of the joints is likely.

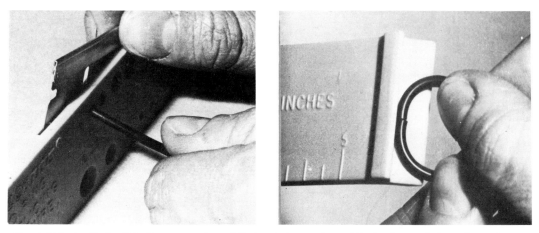

Fig. 4.185 An O-ring kit allows any size of O-ring to be made rapidly in the workshop. To make a ring, first determine the correct size and then cut a piece of suitable diameter rubber to the correct size, using the tool supplied. Apply the glue to the ends of the rubber and use the tool to hold the ring in position whilst the glue sets. Finally cure the join with the water-proofer supplied in the kit.

Once set, this type of sealant is difficult to break.

● *Semi or delayed hardening types*. This category, such as Loctite's Master Gasket and Red Hermetite, is probably the best-known sealant. Master Gasket is anaerobic and will not start to cure until the joint has been bolted together and the air excluded.

Both compounds are excellent for coating gaskets during assembly of engine transmission, pumps, chainsaws and so on. Master Gasket can also be used to make a replacement gasket.

● *Non-hardening types*. Examples like Golden Hermetite and Solvol Blue Hylomar have virtually hundreds of applications.

They form thin but strong seals so are suitable for engine assembly work, especially when the joint surfaces are smooth.

● *RTV silicone compounds*. RTV (Room Temperature Vulcanising) compounds can be used to replace most paper or cork gaskets provided thickness is not critical. They are ideal for repairing or patching a large gasket such as a rocker gasket and can be used in an emergency to make a flat rubber seal or patch up cooling system hoses. Main disadvantages are that normally they cannot withstand high temperatures. Fuel oil and petrol cause them to deteriorate and some of them give off corrosive gases during curing.

There are suitable adhesives for practically every material. There are special ones for vinyl and plastics which actually fuse the edges of the joint together.

One of the latest adhesives is the cyano-acrylate range such as Loctite's Super Glue, particularly useful for joining metals and small rubber components.

The anaerobic types of adhesive which only set in the absence of air are usually used for such specific applications as preventing fasteners working loose or bearings rotating in their housings. Some, such as Loctite's Quick Metal, can fill the gap between a worn bearing and its housing.

Others require ultraviolet light to cure and will therefore set faster on a sunny day.

Anaerobic adhesives are often sold in containers which appear to be half full of air.

Do not attempt to exclude this air by squeezing the container or the adhesive will start to cure.

Fig. 4.186 Use a locking compound to secure a rotating bearing and reduce wear of the housing. Quick-metal will fill a gap up to 0.25 mm if necessary. With the addition of another compound known as Activator N, Quick-metal will fill even larger gaps.

Fig. 4.187 Where a suspect gasket must be re-used, apply one of the semi or delayed hardening compounds to prevent leaks.

Fig. 4.188 Large gaps such as this end section of a power-harrow bed top cover, can be filled using one of the RTV compounds. However, it is advisable to provide support. In this case an additional gasket was bolted in place using angle iron.

Fig. 4.189 Many sealants including RTV compounds can make a replacement gasket by applying them to one of the surfaces to be sealed. Take care not to use excessive amounts of the compound, in case it finds its way inside an engine or transmission assembly. When joints have to be broken without damaging the gasket material, apply a layer of oil or grease to one surface.

Fig. 4.190 Several products are available which are designed to prevent components working loose. They will normally lock and seal nuts and so on without the use of spring washers or other locking devices.

Epoxy adhesives like Araldite are usually sold in two containers. One contains the resin; the other contains the hardener. The two compounds must be mixed in the correct proportions to achieve a strong bond.

SAFETY HINTS

Every year mechanics and farm workers are injured while carrying out routine repairs. The accidents are usually caused by a moment's thoughtlessness, by discarding safety arrangements or, more often, by ignorance of the dangers involved. Safe working requires self-

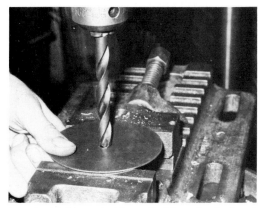

Fig. 4.191 Never hold small objects in the hand when drilling, especially when the article has sharp edges. The drill may pick up the work piece, spin it round like a power saw blade and injure or sever fingers.

discipline. Always wear the right clothes. Never wear overalls with floppy sleeves or a dangling necktie. Foot protection is vital, preferably heavy-duty boots with reinforced toe caps and thick soles.

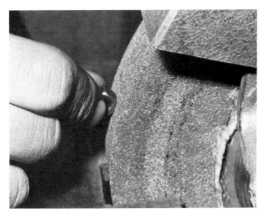

Fig. 4.192 Always use a pair of pliers, or similar strong grips, when grinding small objects. A finger could be badly grazed or trapped. Some small objects could be snatched from the hand and cause a wheel burst.

Fig. 4.193 Do not leave a chuck key in its chuck. If the drill is switched on, the key could easily fly out and into someone's eye. Provide a special place for the key and get into the habit of keeping it there.

Keep the workshop clean and tidy. An untidy working area can cause accidents.

Never store inflammable liquids in unmarked containers and keep such liquids away from heat and sparks. A workshop must have at least one fire extinguisher, preferably of the dry powder or carbon dioxide type, which can be used on fires, involving electrical equipment and flammable liquids.

Adequate eye protection is often neglected.

Fig. 4.194 Always remove expanding metal watch-straps before working near live electrical connections. The strap could bridge a live point, earth it and become red hot in seconds.

Fig. 4.195 Batteries and naked lights or sparks must not be mixed. Never charge a battery near a grinder. The sparks from the grinder could ignite the hydrogen and oxygen gases given off during the charging process.

Fig. 4.196 Cutting or welding near an oil drum is a lethal practice. Oil gives off inflammable vapours when warmed. Never try to weld a tank that has contained an inflammable liquid without first seeking advice.

The correct goggles or face shields should be kept near gas and electric welding plants, grinding machines and drilling machines. Use goggles whenever any chiselling is carried out.

Treat the mains electricity with the respect it deserves. Never overload sockets. Electricity

Fig. 4.197 Never use a pile of bricks or concrete blocks as a support. They are likely to crumble without warning. Never ever crawl under vehicles.

will always seek the easiest path to earth. Using power tools with damaged cables or cables containing makeshift joins is asking for trouble. Should part of your body touch a live wire, the easiest path to earth is through you. Using unsafe power tools in damp conditions is even worse. Current could reach you by tracking along the outside of the power cable, especially when the earth wire is disconnected.

All workshops should carry a first aid box which must be checked periodically to ensure that none of the dressings is missing or that items such as scissors have not been borrowed for other purposes and not returned.

Fig. 4.198 Never remove a wheel and then depend solely on a jack for support. Always support the weight on axle stands or wooden blocks. Axle stands are preferable.

(b)

(a)

(c)

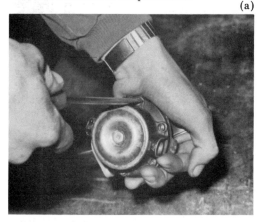

Fig. 4.199 NEVER use an air-line to dust-off clothing. Dirt or air could be forced into the skin with fatal consequences. NEVER use chisels or punches with 'mushroomed' heads. Bits can fly off when the hammer strikes. And — as shown in the illustrations — NEVER use a screwdriver like this (a); NEVER blow dirt off batteries with an air-line (b). It could suck out and blast acid in the eyes. (c) NEVER allow oil or grease and welding gear to mix.

CHAPTER 5
Welding Ways

The fully-equipped workshop has both gas and electric arc equipment. To do in the easiest way all the jobs that crop up could mean a capital outlay of £1000.

It is possible, however, for as little as £200 to £300 — less for second-hand equipment — to obtain a gas or electric welding kit sufficient to tackle most farm workshop jobs. If both gas and electric can be justified so much the better; arc welders are more quickly set up and more easily used for awkward jobs inside and underneath machines, whereas gas is better for high temperature and tricky low temperature work. When it is a choice between the two there are a number of points to be considered.

For the man who is quick to pick up workshop skills the electric type is probably the better bet — particularly where the equipment is needed for welding only. A 150 to 200 amp arc welder — large enough for most farm jobs — can be bought for £200 to £250, the same price as a new gas welding kit.

Gas welding, on the other hand, is more flexible, since it makes possible a number of jobs apart from welding.

Oxyacetylene equipment can be used for

high speed metal cutting and for low temperature work like brazing and bronze welding. It also forms a convenient source of concentrated heat for jobs like tine straightening and light forging (Fig. 5.1).

Electric arc sets can, in some situations, be used for the lighter low temperature work by the use of the carbon arc technique.

The basic technique for welding mild steel with both types of equipment can be learnt by the average tractor driver in a short time at a county college course. On balance, proficiency in gas welding is probably more difficult to achieve than electric arc welding. With either method, practice makes perfect, so stick to the simpler jobs to begin with — someone's life maybe at risk if a job is not properly done.

Equipment needed apart from the welding tackle itself will be special spanners, wire brushes, protective shields or goggles and a leather apron and gloves for the heavy work. For the smaller, lighter jobs a proper welding table is most useful and can be made in the farm workshop (Fig. 5.2).

Fig. 5.2 Essentials for use with gas welding tackle: a flint lighter, valve key, union spanner and large spanner for fitting valve gauges to bottles. Wire brush for cleaning. Goggles, which must always be worn. Leather gloves and apron for protection.

The table is easily made from 38 mm angle iron using the welding kit. For gas welding it should be made to hold 230 x 100 mm firebricks which are much cheaper than the larger blocks normally sold for this purpose, to give a good safe welding and heating surface. If the equipment is second-hand check that the most common nozzles are included (Fig. 5.3).

Fig. 5.1 The front torch of this gas welding kit is the standard tool and will take the full range of nozzles. The torch at the back is for cutting. The large nozzle, far left, is for metal heating.

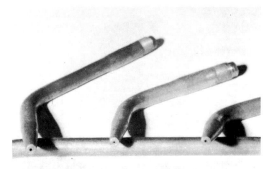

Fig. 5.3 The three most commonly used nozzles are from left to right: No. 25 for welding metal in the 6 to 8 mm thickness range, No. 3 for very thin metal and No. 7 or 9 for 3 mm thick plate.

As a good electrical contact between the job and the earth side of the welder is essential for arc welding, the table top for this type of work should be made of 6 mm plate. The larger job can be done on the fire bricks with the earth lead from the welder attached directly to the job — in the same way that direct work on a machine would be done. Where both types of welding are used the plate top can be removable. At all times, make sure that there are no oily rags, cans of petrol, or other combustible materials in the close vicinity (Fig. 5.4).

Fig. 5.4 This 200 amp a.c. electric arc welding machine gives sufficient power for most farm jobs. The welding table is fitted with a steel plate top and flash screens. For gas welding the plate is replaced by firebricks.

When used for electric welding the table must be fitted with a flash screen — temporary or even permanent blindness is possible from only momentary exposure to arc light at close quarters. The screen must be of fire-proof material stretched over a 25 mm tubing frame which will fit into sockets at the corner of the table.

So, for a capital cost of as little as £200 and the investment of some time in learning how to use the equipment, you are set up to do a whole host of repairs which cost much more when done by outsiders — not counting the cost of the time wasted while waiting for the job to be done. In the next few pages we will explain the basic techniques of both types of welding and the outlines of some more difficult jobs.

GAS WELDING

To ensure easy working and safety, oxyacetylene welding equipment must be handled, set up and adjusted in the correct manner. Oxygen and acetylene bottles must be firmly fixed in an upright position (Fig. 5.5) either to the side of the welding table or in a wheeled trolley stand.

Fig. 5.5

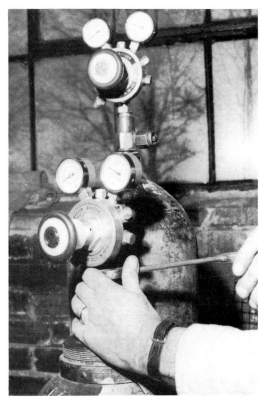

Fig. 5.6

marked 'blow-pipe') are on the torch end (Fig. 5.7). Connect the torch to the other end of the piping — still following the colour coding. Select the correct nozzle for the thickness of metal to be welded (Fig. 5.8) and attach it to the torch. Before tightening the union nut make sure that the end is firmly seated in the torch socket.

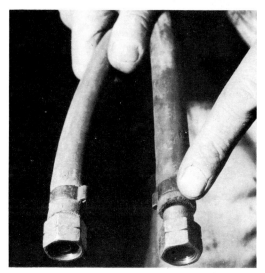

Fig. 5.7

Before fitting the gauges blow the sockets clean by opening and closing the bottle valves quickly; this ensures a gastight seat for the gauge connections (Fig. 5.6). The bottles are colour coded — maroon acetylene, black oxygen. To make doubly sure that there can be no mistakes all acetylene connections are left-hand thread and those on the oxygen right-hand thread.

The gauges are screwed in hand tight and then taken up a further half turn with the correct spanner. Before lighting the welding torch test the system for leaks. With the gauge valves turned off, switch on the bottle valves one turn and set the left-hand working pressure gauges to between 0.2 and 0.3 bar. Note gas pressure in the cylinders on the right-hand gauge, then switch off the bottle valve. If the gauge units are in good order and are properly connected to the bottles, the needles should not move: if they do the seatings are dirty or the units need reconditioning.

Next, connect the piping — it comes in 4.5 m lengths which can be doubled with a connector for jobs a long way from the bottles — making sure that the flame traps (clearly

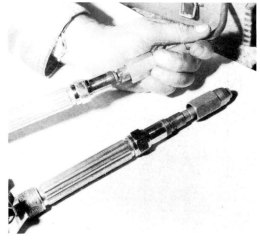

Fig. 5.8

The nozzle must be set at the correct angle for easy working. Hold the torch in the working hand (Fig. 5.9) so that the valve taps are well clear of the inside of the wrist and allow completely free movement, and set the nozzle at an angle of about 60° from the horizontal before tightening the union nut (A) with a spanner.

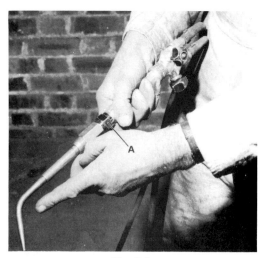

Fig. 5.9

Having ensured that the oxyacetylene welding equipment is correctly set up for easy and safe working, you are now ready to light up (Fig. 5.10).

It is most important to set the flame properly — there must be the right balance of acetylene and oxygen and no gap between the flame and the tip of the nozzle.

With both the gas and the acetylene gauges set at between 0.2 and 0.3 bar (Fig. 5.10(a)), turn on the acetylene and ignite, then turn

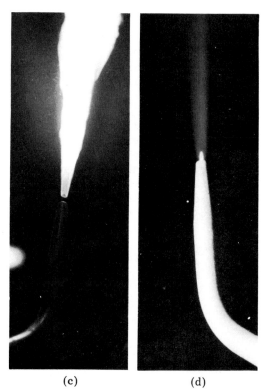

(c)   (d)

up the gas on the torch valve until just above the point at which the flame smokes (Fig. 5.10(b)). The flame must not be separated from the nozzle tip (Fig. 5.10(c)). Now turn on the oxygen slowly until the dark blue and the light blue flame cones coincide to give a fine, sharp, bright 6 to 9 mm 'neutral' flame (Fig. 5.10(d)).

Hold the torch with the nozzle at 60° to 70° to the work and the rod at 30° to 40° (Fig. 5.11). Both must be in line with the weld. The aim is to keep a pool of molten metal under the torch by moving it slightly from side to side to ensure that both sides of the join are being heated and to feed in the weld metal under the flame point by a steady in-and-out movement of the rod.

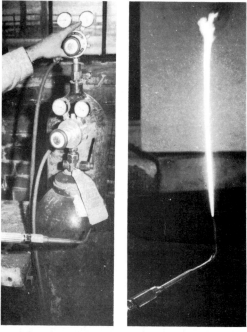

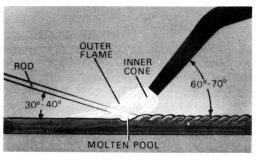

Fig. 5.10 (a)   (b)   Fig. 5.11

Select your welding rod according to the thickness of metal to be welded (Fig. 5.12). The rods (illustrated) are: 1.6 mm for metal up to 2 mm, 2.4 mm for metal up to 6 mm and 3.2 mm for metal up to 12 mm (and above if necessary). The pieces of metal to be welded should be placed the same distance apart as their thickness and tacked at each end by melting the two edges and dipping in the welding rod. Welding proceeds from right to left (Fig. 5.13) and the two edges should be positioned so that there is a slight V shape between them in a leftward direction, to allow for contraction of the joint as the metal gets hot.

Fig. 5.12

Fig. 5.13

ELECTRIC ARC WELDING

Electric arc welding consists in passing a high amperage, low voltage, electric current through the piece of metal to be welded and, by making the current jump or 'arc' from one contact — the welding rod or 'electrode' — to the metal, causing great heat to be concentrated on it. This has the same effect as a gas welding flame but, because the welding rod is part of the apparatus, only one hand is needed for the job.

The electric arc forms a pool of molten metal in exactly the same way as a gas flame, and the molten metal which fuses with the two edges of the job comes from the combined welding rod and 'torch'. There is a risk of contamination of the molten metal — which could weaken the joint — by some components of atmospheric air, so the welding rod is protected by a coating of flux which melts with the metal and forms another protective cover on the weld until it cools (Fig. 5.14). This solidified cover hardens and is easily chipped off with a slag hammer when the job is completed.

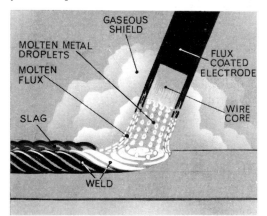

Fig. 5.14 Electric welding proceeds from left to right and the flux surrounding the welding rod forms a protective cover on the metal as it cools. This is 'slag' chipped off when the job is finished.

The basic arc welding set-up (Fig. 5.15) includes the earth side of the welder which is connected to the metal top of the table by the spring-loaded earth strap (extreme right), and the welding 'torch,' the hand grip (right centre). The eyes are protected from the fierce glare by either the hand-held shield (rear centre) or the head clip type shield (rear right). The slag hammer (left centre)

Fig. 5.15

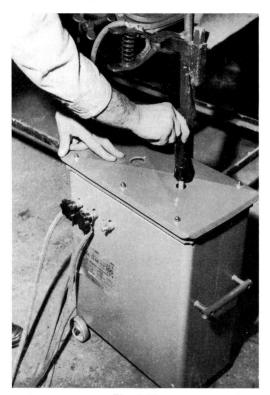

Fig. 5.16

safer. High voltage reduces the amperage range but allows easier maintenance of the arc and higher heat at any given amperage. Finer amperage adjustments are made with the handle, which registers the set amperage on the dial (by the operator's left-hand fingers). The amperage is set according to the thickness of the welding rod (or 'electrode'); typical examples are 30—60 amps for 1.6 mm, 90—100 amps for 3.2 mm and 120—140 amps for 5 mm. Correct amperages are marked on welding rod packets.

The thickness of the electrode or welding rod is selected according to the thickness of metal to be welded (Fig. 5.17). Typical examples would be 3.2 mm rods for 6 to 7 mm thick metal and 4.0 mm rods for 9 to 12 mm thick metal.

Fig. 5.17

is used to chip off the deposit left by the flux on the weld and the wire brush must be used to clean the joint before welding.

There are usually two alternative voltage connections on the welding unit (Fig. 5.16). Low voltage allows higher amperage and is

The rod is held at an angle of 80° and the hand must be lowered towards the work as the welding rod or electrode melts away, to maintain the tip at the same distance from the work (Fig. 5.18). The angle of the rod is increased as the thickness of the rod increases. Otherwise the operation is the same.

Before starting work set the welder to the

Fig. 5.18

correct amperage for the size of the welding rod in use. The two pieces of metal are tacked together as for gas welding and the welding proceeds from left to right.

Brace the right-hand side of the body against the side of the welding table, to help steady the hand. Strike an arc by touching the end of the rod quickly on the work — the action should be like that of striking a match. Draw off and maintain the rod 2 or 3 mm from the work. This needs practice; you will weld the electrode to the work several times before you get the knack. When this happens remove the welding handle quickly from the rod and knock if off with the slag hammer.

A steady hand is probably more important with electric than with gas since the electrode must be kept steadily at the right distance from the metal as it burns away.

Before the arc is struck the eye-shield must be moved in front of the eyes; never strike an arc with the eyes uncovered, as painful temporary blindness can result. This precaution also applies to helpers and onlookers.

MAKING A TABLE

To show oxyacetylene welding equipment being used for a practical job, we show the step by step construction of a welding table (Fig. 5.19).

Fig. 5.19

The main materials are 32 mm steel angle iron for the frame and 32 mm T-section for the internal supports of the firebrick table top.

After cutting to length, the main job is welding up the table top. Simplest way of joining the corners is a butt joint (Fig. 5.20). A 45° join is neater but more difficult. Before

Fig. 5.20

Fig. 5.21

the joint can be welded the edges must be ground back to give a 60° included angle between the two edges (Fig. 5.21) to ensure that the weld penetrates right into the joint.

The joint is welded on the underside to leave the top flush for the bricks to seat properly. Lay the two pieces — held upright by clamps — on a bench and make sure that they are at right angles (Fig. 5.22) before welding.

Fig. 5.22

The secret of a job like this is to have sufficient heat to warm up the metal quickly and to be able to work quickly without burning the metal — so use a no. 10 nozzle.

To prevent expansion distorting the joint, the two edges of the weld must be 'tacked' before starting the weld proper. But before tacking, heat up the metal for a few seconds with a neutral flame (Fig. 5.23) this will also be used for the job. Get a molten pool at the

Fig. 5.23

Fig. 5.24

edge and then just dab in the welding rod briefly and then, after re-checking with the set square, repeat the process on the other edge of the joint (Fig. 5.24).

The welding then proceeds (Fig. 5.25) following the drill shown. The rest of the table top is then welded up by joining the T-section pieces at 25 cm centres into the frame. Straight butt welds are used to attach the legs — with regular checking with the set square — to the table top. A frame similar to the top, but 12 mm shorter and narrower is constructed for the leg stays and is welded into the angle of the legs. Fitting the firebricks into the channels completes the job.

Fig. 5.25

Welding Table Cutting List

Two pieces 32 mm angle	610 mm long
Two pieces 32 mm angle	1,210 mm long
Two pieces 32 mm angle	595 mm long
Two pieces 32 mm angle	1,190 mm long
Four pieces 32 mm T-section	610 mm long
Four pieces 32 mm angle	760 mm long

Arc welding

Cutting-out details and sizes of the parts to make a table are the same for arc welding as shown in the article on gas welding. Joints can be either 45° angle or square. The latter — shown in Figs. 5.26 to 5.30 — are easier to handle and weld.

Greater penetration power of the electric arc means that preparation of joints by grinding an angle, as with oxyacetylene, is unnecessary.

Clamp one side of the joint to the bench and tack on one side only. Then check with the set square (Fig. 5.26) and tack on the other side. Now weld the vertical corner of the joint (Fig. 5.27) fixing in an upright position with the clamp. The completed joint here shows one of the points which must be watched

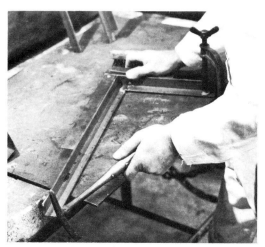

Fig. 5.26

Fig. 5.27

Turn the joint back and reclamp to the table for the main weld (Fig. 5.28) taking care to clean the metal before starting. A 3.2 mm electrode using about 125 amps is used to ensure penetration into the joint. A steady, fairly fast action round the angle of the joint (Fig. 5.29) will avoid burning the metal.

Fig. 5.29

Aim for the sort of weld shown in Fig. 5.30 with the slag tapped off. This standard can be attained only with practice.

Fig. 5.30

in arc welding; there is a deep flaw or 'blowhole' in the weld which will seriously weaken the joint. This must be ground back with a portable grinder before rewelding.

Fig. 5.28

WELDING JIG

A simple welding jig for ensuring the right-angled joints are welded up squarely comes from Mr I. A. Brown. It consists of two pieces of 38 mm x 38 mm box section 450 mm long welded to form a 90° corner (Fig. 5.31).

When two pieces of metal, either box section

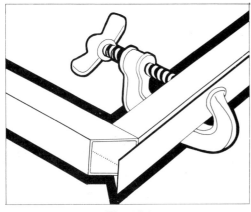

Fig. 5.31

or angle iron, have to be welded at right angles, Mr Brown clamps them to the jig so that his hands are free for welding. He mitres the ends of angle iron before clamping the joint. The corner of the jig is cut off diagonally so that he can make the inside vertical weld while the work is on the jig. When the outside vertical weld and main weld are done, he removes the work from the jig and completes the main internal weld.

For box section joints Mr Brown clamps work in the same way as for the angle iron, but completes the weld before removal from the square.

Apart from the obvious advantage of lining up the joint quickly, the job can be left to cool on the jig and cannot be pulled out of square as usually happens when free-hand welding cools down and distorts due to contraction.

SLEEVE WELDING *(Fig. 5.32)*

Where lateral strengthening is needed in structural work such as cattle pens or farm gates the joints can be welded with a rod or tube insert fitted. But for crop sprayer booms and other liquid carrying pipe work the flow must not be restricted by weld penetration or insert supports.

Sleeving combines lateral strengthening with an effective repair. Where pipes have been fractured and the broken ends cut off, a sleeve will bridge the gap without shortening the pipe.

Welders with marginal experience will appreciate that a strong watertight joint in two pipes of equal diameter can be difficult to achieve.

Fig. 5.32 Cut a sleeve to make a close fit over the pipe and overlap the break by 50 to 75 mm.

The arc process is used here with a 2.6 mm general purpose type mild steel electrode and the current set at 30 amps.

A small overlap of weld will help to seal the join. The pipes used in this case were 25 mm galvanised steel water pipe sleeved with similar 30 mm diameter pipe.

In a fractured steel water pipe the first job is to saw off the broken ends and file them smooth. If the work has been dismantled it should first be measured so that it can be spaced in the sleeve to its original specification.

Cut a piece of tube to fit the pipe snugly and to a length that allows 50 to 75 mm of overlap on each side. Tack in position with a spot of weld if there is any difficulty in supporting it for welding. Weld in a vertical position as this will make easier the continuous weld that is essential for good work. The weld should be built up to overlap by 6—12 mm to ensure a good watertight weld. Gas or electric arc welding can be used for this type of work.

BRONZE WELDING *(Fig. 5.33)*

Bronze welding is quite different from the fusion welding process, though the same basic equipment is used. The bronze rod is melted on the hot metal and flows into the small pores of the metal by capillary action to give a strong bond between parent metal and rod deposit.

(c)

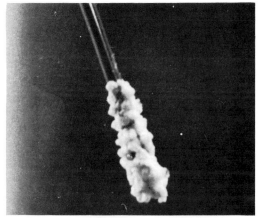

Fig. 5.33 (a)

(d)

(Pictures by courtesy of Rycote Wood College, Thame.)

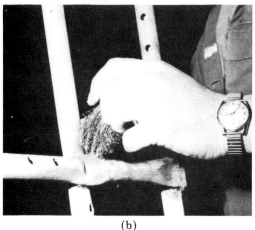
(b)

Its main advantage is the low temperature — about 850°C — at which it is carried out instead of 1200°C plus for ordinary fusion welding. The low heat input into the parent metal gives less distortion than ordinary welding and is useful in joining thin sheet metal.

Bronze welding can be used to join dissimilar

metals, such as copper and steel, which cannot be fusion welded because of the wide difference in their melting points. Repair of malleable castings is best done with bronze welding, as fusion welding's high temperature will destory the properties of the casting.

A borax-based flux is used in bronze welding. This cleans the surface of the metal to enable good bonding of the bronze on to the parent metal and helps float off impurities. Special flux-coated rods can be obtained but are expensive.

Flux can be applied by dipping the rod into the flux (a) or it can be mixed into a paste with water and brushed on to the joint before welding commences.

Rods most commonly used are nickel bronze for high strength joints or dissimilar metals, manganese bronze for malleable castings and grey cast-iron, and silicon bronze for mild steel or galvanised materials.

Use an oxidising flame to prevent the zinc being boiled away. Insufficient oxygen gives a porous weld. Cleanliness of the joint (b) is essential for a good bond to be formed between the bronze and the parent metal. Clean off all grease, paint and rust and roughen the surface of the metal with a file before welding. Remove all burrs and sharp edges or they will overheat and spoil the weld.

In the example shown, 25 mm diameter galvanised pipe is being bronze welded using 3 mm diameter silicon bronze rod. Gas pressures are 0.15 bar for both bottles and a size 7 nozzle is used. Heat the metal to dull red and apply the rod, which should melt if the parent metal is at the correct temperature and leave a deposit which can be built up to the required size (c). Avoid overheating the parent metal or the tinning action of the flux is spoilt and the rod deposits run into beads.

The finished job should have a smooth deposit, free from air holes (d).

Zinc fumes are poisonous, so carry out bronze welding in a well-ventilated area.

JOINING ROUND BARS

Welding together mild steel rods (Figs. 5.34 to 5.36) for shafting or construction work is a fairly straightforward job for the average farm workshop operator. It is also the type of work that can have a number of applications on the farm and a knowledge of the correct procedure will help to avoid the poor or unsightly work that can result from lack of attention to detail.

The oxyacetylene method is illustrated

Fig. 5.34 The ends of the two sections of 16 mm diameter mild steel are prepared so that they are bevelled to a double vee at an angle of 45 to 60°. Wear on the grinding wheel will be reduced if the outline is first cut with a hacksaw.

Fig. 5.35 The two ends are placed on the weld table with a gap between them of 1.5 to 3.0 mm. The work is heated until it is thoroughly molten before using the 1.5 mm filler rod to complete a run as if welding plate, i.e. with good penetration. The work is then turned over and the run built up — it is essential that a good weld is achieved at this stage.

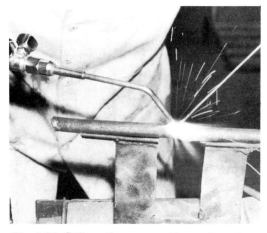

Fig. 5.36 Still on the reverse side continue to build up the join using the 3 mm rod. Do not weld one run on each side alternately or flaky slag will develop and hamper proper work. When this is completed the build-up continues on the other side finishing with a few runs lengthways across the join to leave a good build-up of weld material.

here using 2 and 3 mm diameter rods respectively with No. 7 and No. 10 nozzles. The oxygen pressure should be set at 0.3 bar, the acetylene at 0.2 bar and the flame adjusted until it is neutral.

HARD SURFACING

At today's prices and replacement costs, hardsurfacing of wearing metal parts in machinery is obviously useful but the techniques were difficult in the past. There are now two new techniques which make hardsurfacing easier, more accurate and cheaper.

SPRAY WELDING *(Fig. 5.37)*

It is worth taking a look at spray welding. This is not an entirely new technique but modern equipment has improved it significantly. The idea is to use a specially designed gas welding torch to spray metallic powder on to the surface of a metal part where the flame produces a fused hard surface. A conventional torch with provision to add a can of metallic powder on the top of it and a trigger to release a flow of powder into the flame can be used. The powder sprayed on to the surface of the workpiece is melted by the flame and fuses with the component material. It is a very useful process for surface hardening of cultivator points, plough share discs and so on. It can also be used for building up worn parts. The spray deposit can be machined so that, with the choice of suitable powder, the technique can be used for building up worn shafts which can be turned on a lathe back to the original design size.

One of the major advantages of the technique is that it is very easy to operate. Even people who have not welded before can learn the basics in about ten minutes. The normal welder of average ability will be able to produce good deposits almost immediately and be able to control the evenness and thickness of the added layer. The whole operation is much easier and faster than hardsurfacing by conventional methods of gas or arc welding.

For spray welding, you need really clean, shiny metal parts. De-greasing of clean parts with a steam cleaner is good practice. Even finger-marks are best kept off the clean surface and spray welding should be carried out as soon as possible after cleaning, before surface rusting occurs.

The tip of the welding pistol should be held 15 cm or so from the work. The hardsurfacing is best applied in thin layers of about 0.254 mm to build up an 'overlay' of powder material on the metal to be surfaced. Several layers can be applied to build up a thicker layer. Attention should be paid to getting an even deposit. After application of the overlay, the temperature of the metal is raised with the welding torch only to a dull red heat. As the overlay melts and fuses to the workpiece surface it will develop a shiny, mirrored surface, and this glazed pool can be drawn along the metal, taking care not to overheat.

Spray welding is not yet a common farm

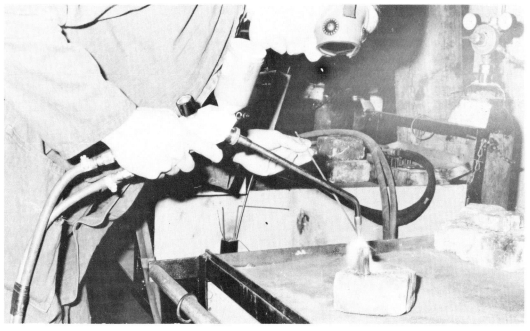

Fig. 5.37 Spray welding showing the gun with powder cannister.

workshop operation. It has often proved to be more expensive than traditional hard facing methods in many instances.

Alternatively, there is a process that does not need the special pistol for applying powders. You could save the £100 on the kit. The process uses an ordinary torch and a special paste which is 'sweated' on to the metal.

Wall Colmonoy Ltd (of Pontadawe, Glamorgan, Wales, UK) have developed sweat-on paste for hardsurfacing. It is based on chromium boride crystals which are, in the company's words, 'second only to diamonds on the hardness scale and the most abrasion resistant material available for hardsurfacing'. (Fig. 5.38).

Fig. 5.38 Sweat-on paste being used with a standard gas torch.

There are three ways to apply sweat-on paste. The gas welding technique is recommended for thin metal sections, the carbon arc method for thicker sections, and the gas tungsten arc technique for smoother, less diluted deposits.

GAS WELDING

This method, employing an oxyacetylene torch, is ideal for metal thicknesses up to 1.5 mm, and is also very useful for plough shares, harrow discs, cutters and similar farm implements, coal cutter bits or excavator teeth.

To begin, the area to be handsurfaced must be clean and freed of scale or rust, by grinding if necessary. Paste must be mixed thoroughly, *adding water if necessary*, to bring it to a brushing consistency.

Apply a thin overlay of paste, approximately 2 to 3 mm, to the surface with a fine bristle brush to provide complete coverage. *Allow to dry.* Then, using a reducing flame with feather equal to the length of a cone (a 2X flame), begin fusing the paste to the base metal with an oxyacetylene torch. Heat should first be applied to the bare base metal adjacent to the overlay of paste. When this metal becomes red, apply the flame to the paste surface until the paste sweats into the surface of the steel. Sweating is indicated by the quenching of the bright colour as it bonds to the base metal. Heating should progress as fast as the sweating to prevent puddling of the steel surface.

Carbon arc welding, and gas tungsten arc equipment is not normally available on the farm but it is preferred for heavier work, it would be better for heavier earth engaging parts such as plough shares or chisel points but the use of normal gas welding equipment is ideal for thinner components such as discs and can give good results on the thicker and heavier parts.

NOZZLE CARE

Gas welding tips and nozzles should be stored in a rack or clean box to prevent the gas hole from becoming blocked by dirt. When fitting a nozzle to the torch use the correct spanner. Pliers will damage the flats on the nut. Tighten the nozzle just enough to provide a gas-tight joint.

Over a period the gas hole will become misshapen due to heat from the metal being welded, blowbacks, metal spitting back through overheating, or the nozzle touching the molten welding pool. An odd flame shape difficult to control will form and give an uneven weld (Fig. 5.39).

Fig. 5.39 A worn welding nozzle with a poorly-shaped gas hole, is shown alongside three normal nozzles.

Clean the tip by using a fine file until the tip is square and smooth (Fig. 5.40). To reshape the hole use a set of nozzle reamers. These are sold in sets of 14, which will cover most standard nozzle sizes. Each reamer is made of hardened metal with fine cutting teeth on the shank.

Fig. 5.40 File the nozzle end until perfectly flat.

First use a reamer a size smaller than the nozzle size, working it in and out of the hole, removing the burrs and carbon particles. Do not overdo it or you will have an oversize hole that will not deliver the proper volume of

gas at the set pressure and will not function properly. Then change to the recommended size of reamer and repeat the operation. If no reamers are available a small drill will do for the larger sized nozzles, but do not twist it or it will cut the soft copper and enlarge the hole (Fig. 5.41).

Fig. 5.41 The final operation is to ream out the hole.

Light up after cleaning and check flame shape. Welding flames should be smooth and straight. Pre-heat flames on cutting nozzles should be of equal size and straight, with the oxygen jet evenly shaped and smooth.

Carbon deposits inside the nozzles and shanks can be removed with a powder from the manufacturers of the welding equipment. Dissolve a small amount of powder in water and soak the nozzles for 24 hours to loosen the carbon particles. Dry off the nozzles and blow them out with an airline to remove all the bits.

On heavy work, where a large nozzle producing a lot of heat is used, the nozzle and shank will get hot from reflected heat. To prevent overheating plunge the torch into cold water occasionally during work. Shut off the acetylene, but leave a small volume of oxygen flowing to prevent water entering the torch valves.

NOZZLE RACK

Store welding nozzles and spanners on the wall by the welding bay, or where the welding equipment is kept. This easily-made rack for nozzles consists of a batten of wood with holes drilled for each size nozzle and the number painted by the appropriate hole (Fig. 5.42).

Fig. 5.42 Welding nozzle rack.

For a spanner rack, paint the outline of each spanner on a length of wood and knock in nails to hold each one in place. You can then see at a glance if a spanner is missing or in the wrong place.

ARC WELDING IN ACTION
(Fig. 5.43)

Suppose one end of the yoke on an M-F tractor radius rod snaps off after collision. A new arm will cost several pounds, but a welding repair, just as strong, takes only an hour.

First prepare the broken ends for welding. The metal is about 10 mm thick so use a double-vee preparation to ensure good fusion of the two parts throughout the whole thickness of metal. Angle the broken edges at 45–60° on a grinding wheel. Do not take any more metal off than is necessary or a big gap will be left between the two parts that will be difficult to fill with weld deposit.

Next task is to set the parts in their correct position so that when rewelded the job is as near specification as possible. It must be lined

Fig. 5.43 (a) Clean the break, clamp the arm then line up the arms and the holes.

(b) Check progress after each run, and chip away slag.

up two ways. First, the holes must be in line with each other, and secondly the distance between the two arms must be the same as before.

Clamp the arm in a vice and use spacers and washers, and a length of bar through the yoke eyes, to line up the arms. Clamp the job rigid with G clamps.

For an arm made of medium carbon steel

(c) Add a bead along both sides of the weld for added strength.

ideally a nickel and steel alloy electrode with low heat input should be used to minimise distortion. It leaves a malleable deposit that work-hardens, but the electrode is expensive.

A cheaper alternative is mild steel electrode 3.2 mm at 90 amps. Three runs might be needed on each side of the vee. Check after each run that the job is still true as the welding heat may pull the parts out of true. To give extra strength to the job, lay a bead along the edge of each side of the yoke. After welding, peen the weld with a hammer to relieve any stresses that have built up.

Fig. 5.44 A length of angle iron helps ensure that the cut is straight and at the correct angle for welding. Clean up rough edges on a grindstone.

WELDING THICK METAL

There is no need for special preparation when welding material up to 5 mm thick, except to remove any rust and to leave a narrow gap between the two pieces to be welded to ensure adequate penetration.

Where material is more than 6 or 7 mm thick electric arc welding is generally used for speed and ease, and the material must be 'veed' out to an angle of 45—60° before welding commences. 'Veeing' gives good penetration of the arc and allows the metal to be fused throughout its thickness, thus giving a strong joint.

Single vee preparation is used on materials 5 to 10 mm thick where it is difficult to get

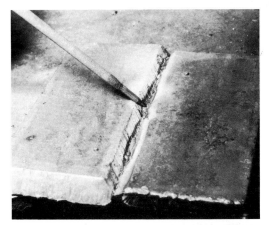

Fig. 5.45 After laying the first bead, turn the job over and check that weld metal has penetrated right through the vee. Here the pointer indicates good penetration, but there are portions where there is little penetration.

Fig. 5.46 After the root runs lay beads on alternate sides of the joint so minimising distortion. The second and third runs are laid on opposite sides of the vee faces and the fourth one along the top to give a level finish.

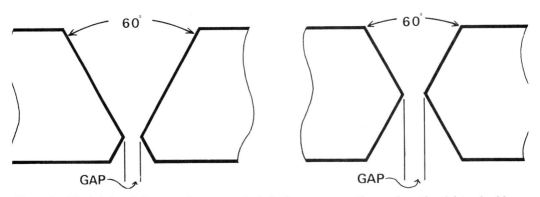

Fig. 5.47 The left-hand diagram shows a typical single-vee preparation and on the right a double-vee preparation. The gap between the two pieces of metal will vary but a rough guide is about 2 to 3 mm.

at the underside of the weld. It requires more rods to fill the vee, is more expensive, and causes greater distortion. Use double vee preparation if it is possible to reach both sides of the weld. Double vee takes longer to prepare but saves electrodes and minimises distortion by laying a run on alternate sides, each bead counteracting the distortion of its predecessor. The best way to make the vee is to use an oxyacetylene cutter and a piece of angle iron. Lay the angle iron on its two edges near the metal edge to be prepared. Lay the nozzle parallel to the side of the angle iron and cut off the unwanted metal at a 45° angle set by the side of the angle iron. You should get a good enough finish to the metal without further preparation, but if necessary clean up the cut on the grindstone and remove slag (Fig. 5.44).

Other methods of forming the vee are by portable grinder, hammer and chisel, or filing though these are much slower.

To weld the pieces together use the largest electrode possible for speed and to minimise distortion, but make sure you do not overheat the metal. First weld the two root runs in the bottom of the vee (Fig. 5.45). Next lay a run slightly to one side of the vee to build up the gap. The next run will be on the other face of the vee and so on, until the weld is up to the level of the parent metal. Take care to remove all the slag for each run, to prevent porosity and poor welds. The last run on the top of the metal should be done with a weave to give a good finish to the weld and strength and reinforcement to the joint (Fig. 5.46).

In the example shown, using 12 mm thick material, the root runs were carried out with 3.2 mm rods at 120 amps setting and the other runs with a 4mm rod at about 145 amps.

CAST-IRON WELDING

Cast-iron is one of the most commonly used metals in agricultural machinery. It is cheap to produce, hard wearing and has a long life. Its drawback is that unless it is malleable cast-iron, it is brittle and easily broken.

The sprocket in the picture was from a combine and was broken by a hammer used to remove it. Fusion welding was used to repair the broken part, though bronze welding could equally well have been used (Figs. 5.48–5.51).

Fig. 5.48 Pre-heating and post-heating should take place at the chalk marks on the sprocket rim to prevent cracking as the metal expands and contracts due to the heat of the welding process. Note the fire bricks round the sprocket to prevent draughts which will cause uneven heating and cooling.

Fig. 5.49 Preparing the crack. A spatula made out of ordinary mild steel welding rod is used to scrape molten metal away from the crack. A 90° vee should be formed to about half the thickness of the casting.

Fig. 5.51 After both sides have been welded clean off the slag with a wire brush, and paint it while still warm. Check when the sprocket is refitted to the machine that it runs true.

Fig. 5.50 Welding the metal. Heat the part to be welded until it is plastic. Test for this with rod using a stirring motion to melt the rod in the molten pool of the parent metal. Two runs were used here, one a root run in the bottom of the vee and a filing run to give a level surface.

The two main points of the job are cleanliness and pre-heating of the sprocket before welding and post-heating after welding.

All grease and dirt should be removed with a proprietary degreasing compound. If petrol or paraffin are used make sure the sprocket is dry before welding. Removing dirt cannot be overemphasised. Carbon from oil or grease can cause gas bubbles in the weld giving a weak porous joint.

Pre-heating to a dull red heat (550—660°C) prevents stresses in the casting either before or after welding. These stresses can cause cracking in other parts of the casting or distortion. Post-heating ensures that the weld cools slowly and a strong machineable deposit is made. Rapid cooling will leave a hard brittle white iron deposit that is difficult to machine.

In the job shown 4.5 mm square rods containing enough silicon to leave a machineable deposit were used. A No. 5 nozzle was used to weld, and both gas bottles were set at 0.2 bar.

A cast-iron flux containing slag producing compounds was used to dissolve oxides of iron and protect the metal from oxidation during welding .

One side of the sprocket was prepared and welded and the process was repeated on the other side. Total time taken was about one hour and the combine went back to work instead of waiting for a spare from the factory.

SPARK LIGHTER

An old magneto mounted on a plate fitted with a set of contact points — gap about 3 mm — or a spark plug is useful for lighting a welding torch. Flick the drive coupling to produce a spark. Weld it to the bottle frame and it is always available for use (Fig. 5.52).

Fig. 5.52

OXYACETYLENE CUTTING

As well as being an extremely useful tool for repair jobs and fabricating equipment, your oxyacetylene welding kit can also be used for fast, moderately-accurate cutting-out of sheet and bar steel. For an additional outlay for a Sapphire cutting head you can tackle most of the cutting jobs likely to occur in the farm workshop.

For these jobs your standard regulator gauges are quite adequate, although for very heavy cutting — metal greater than 25 mm thick — you will need an additional high-pressure gauge set.

The cutting head (Fig. 5.53) screws on to the hand-grip base of the normal welding torch and consists of the special right-angled head, the tube arrangement to provide large quantities of oxygen — which produces the great heat needed to burn through metal — and the special, thumb-grip oxygen control.

Fig. 5.53

With the cutter you should have a set of cutting nozzles for different metal thicknesses.

As with welding, some practice will be needed before you can cut accurately, particularly on curved work. Before starting, mark out the metal (Fig. 5.54) with a chalk line and centre-punch along it. The chalk will disappear as soon as the flame touches it.

Fig. 5.54

Next set the flame. Light up as for normal welding and, using the valves on the torch (Fig. 5.55) — do not touch the thumb valve at this stage — adjust for a normal, neutral flame with an equal balance of acetylene and oxygen.

Fig. 5.55

The only setting difference on the bottle gauges is that the oxygen pressure will be much higher — 1.7 bar for cutting 6 mm metal and 2.1 bar for 12 mm.

The metal to be cut should be held steady by weights or a vice and the body of the torch should be steadied on the fingers of the left hand resting on the bench. Make a swift practice run across the line of the cut, with a little pressure on the excess oxygen thumb-

Fig. 5.56

Fig. 5.57

Fig. 5.58

valve to clean the metal then start to cut (Fig. 5.56).

The nozzle must be vertical and the point of the flame 3 mm above the metal. Heat at the edge until the metal is bright red, then switch on the excess oxygen to set fire to the metal. Draw the torch steadily towards you until the job is completed. Where a series of pieces of metal of a similar length or width are being cut, a guide fitted to the nozzle base (Fig. 5.57) will speed up work considerably.

Another ancillary use for the welding kit is for metal heating — particularly useful for in situ forging jobs (Fig. 5.58). A simple, straight-heating torch is screwed on to the handgrip and, with gauge pressures set at 0.15 to 0.2 bar both acetylene and oxygen, a slightly carburising flame — with more acetylene than oxygen — is set to give concentrated, steady heat wherever it is needed.

CURVE CUTTING

Cutting large holes, circular discs or even curved shapes from mild steel is made easier by a simple jig to hold the cutting nozzle and control its movement to the desired radius.

The one illustrated in Fig. 5.59 was made in a few minutes by Mr J. R. Marshall from a length of 10 mm diameter bar to form a radius piece, with a small square of 6 mm plate welded to the end.

Drill a hole in the plate big enough to take the cutting nozzle on the large end of the taper. A short length of 12 mm bar, drilled through to take the radius piece and on one side to take, after drilling and tapping, a 6

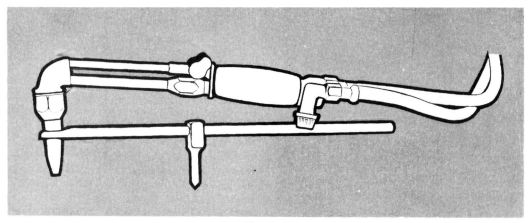

Fig. 5.59

mm diameter stud with a wing-nut attached, forms the body of the pivot point. Weld a short-pointed peg in the bottom of the block to form the pivot.

Since cutting nozzles vary in external diameter and length, the size of the hole in which the nozzle fits and the length of the pivot peg (which will control the height of the nozzle tip above the metal) must be made to fit the actual torch with which the jig is to be used.

A little practice with the jig will enable quite acceptable curved cuts to be made, as long as the basic rule that the cut is always started in the waste metal is observed.

PORTABLE WELDING TROLLEY

In many cases of machine breakdown on the farm, gas welding equipment has to be taken to it, to do a repair. This involves the inconvenience of wheeling the trolley long distances or heavy work lifting it onto a trailer for transport.

Mr Mackay, Head of the Machinery Department at Berkshire College of Agriculture, fits the welding bottle trolley to the three-point linkage of a tractor. He has welded brackets to the trolley, so that it can be attached to a tractor in a matter of minutes. This allows the gas welding set to be taken to all parts of the farm and out to the fields if necessary, to carry out emergency repairs to stranded machinery. While the trolley is mounted on the tractor, the check chains must be tightened to prevent swaying; the gas must be turned off, and the gauges removed from the bottles as any jarring may reduce their accuracy (see Fig. 5.60).

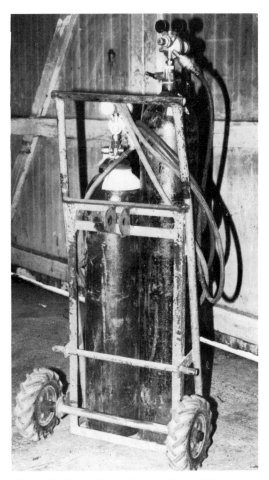

Fig. 5.60 Portable welding trolley with tractor mounting points.

CHAPTER 6
Check-up and Set

PLOUGH OVERHAUL

Beams and frames

A poorly adjusted or incorrectly mounted plough results in bad work and wasted tractor power. Fuel consumption per hectare increases, the cost of replacing wearing parts goes up and work output drops.

A good commercial job of ploughing combines speed with quality of work leaving uniform furrows with a clean level surface free from weed and trash. There must also be plenty of soil available for discs and harrows to get to grips with during subsequent cultivations. For this the plough, like all implements, needs proper maintenance.

Take advantage of a wet spell to do a complete overhaul.

Equipped with an average tool kit plus a straight edge ruler and a plumb line, it can be done in a day. To make things easier and save barked knuckles, give all nuts and bolts a liberal dose of penetrating oil a couple of days before.

All measurements should be taken with the tractor and plough standing on a level surface and to obtain greater accuracy it is wise to fit a new set of shares.

Field setting problems which arise from bent or twisted beams or main frame members may show up in work as uneven furrows, varying depth of work, or, probably the worst fault of all — crabbing sideways.

Fig. 6.1 Use a spirit level and a straight edge to detect main frame distortions. A steel bar clamped across the frame while the bolts are slack may pull a twisted plough back into shape.

First carry out checks for distorted main-frame members and beams. Distortion is chiefly vertical or lateral. Vertical distortion is checked by measuring from under the share point to the under edge of the beam. The distance should be the same for all bodies. Small discrepancies can sometimes be corrected by individual pitch adjusters if fitted; otherwise the share of the body must be shimmed (Fig. 6.1).

When a beam is distorted laterally the share will be taking too much land or not enough. It is not always possible to detect by eye a bent beam; discrepancies will become obvious if one of the three following methods is used.

(1) Place a straight edge along the landside and share point of each body in turn and draw extended lines for a distance of at least 1 metre remembering to allow for the longer rear landside. The distance between the extended lines should be the same for all bodies (Fig. 6.2).
(2) Place a straight edge along the share points (Fig. 6.3).
(3) Drop a plumb line down the face of the beam directly above the share point and mark the place where it contacts the share. The marks should all be in the same position in relation to the edge of the share.

Fig. 6.2 Checking the position of the bodies relative to each other may be done by extending the line of each landside forward on a level surface. These lines should be parallel and an equal distance apart. A discrepancy of more than about 10 mm in 1 m (½ in. in 4 ft) indicates a bent or twisted beam.

Fig. 6.3 An easier way of checking the relative position of bodies — but not as accurate as the extended line method — is to place a straight edge along the share points.

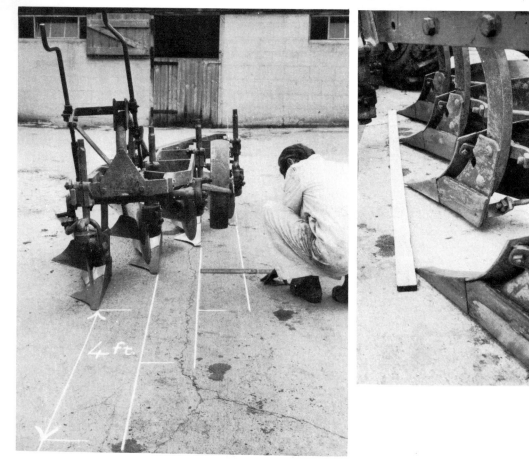

Mouldboards

Having got the main frame right, check on body alignment, disc coulter and skim settings. Making sure, too, that the hitch, draught line and tractor wheel settings are right.

The nose of a body sometimes becomes bent; to detect this, place a straight edge along the underside of the share and landside. The gap at (A), known as the suck, should be the same for each body. Metal packing strips for pegged-on shares or a flattened cigarette packet for bolt-on shares can often be used to correct matters (Fig. 6.4).

The rear mouldboard on this plough is 20 mm too low. They should all be the same height and the distance between each one equal.

Moving the mouldboards in or out also alters the width of the furrow trench. It is an adjustment which comes in handy when using a tractor with oversized tyres to prevent the last furrow from being squashed.

Setting up the reversible plough is often looked upon as being much easier than the conventional types, but this is not necessarily true. Although the settings suggested for conventional ploughs apply equally to reversibles, there are a few extra points to watch. It is essential that both tractor tyres are at equal pressure and lift links must be exactly the same length. This is to ensure even depth front furrows in both directions, and the main frame must be parallel to the ground at all times to achieve level work. Conventional

Fig. 6.4

Fig. 6.5

Fig. 6.6 Distance A should be the same on each body.

This is more difficult to correct and to spot when buying second-hand (Fig. 6.5).

On ploughs with adjustable mouldboards care must be taken not to push them out too far, because this will increase the draught load and may cause the furrows to roll back, especially when ploughing grassland (Fig. 6.6).

ploughs may be tilted in certain conditions to help turn the furrows.

Disc coulters and skims

The initial setting for disc coulters is usually 20 mm or one finger width from the edge of the shares to the discs and cutting just deep enough to leave a clean furrow wall (Fig. 6.7).

Fig. 6.7 Final disc adjustment can only be done in the field. To make this easier, set them all to the same measurements during the overhaul. (A) 50 mm, (B) 15 to 20 mm, (C) centre of disc just behind the share point.

Discs set too deep may prevent the plough penetrating and cause excessive wear in the bearings. Coulter stops should be set to give equal swing on either side of the share. Normal fore and aft setting is with the centre of the discs slightly behind the point to help penetration. Discs should never be angled to land as this puts a tremendous strain on the bearing; angling should be towards the ploughing, making sure they are all equal.

Skim coulter setting for average conditions is just deep enough to ensure that all weeds and trash are buried. Set the point of the skimmers as close to the discs as possible without actually touching them and the heels slightly farther away (Fig. 6.8).

Fig. 6.8 With skims at front of the discs, there is less chance of trash build-up at the back of the skim and skimmings are cut before the furrows become loose.

If they are set too deep or at the wrong angle, the skimmings do not go into the furrow bottom, causing hollows under the furrows, and there is a danger that they will be dragged on to the top during subsequent cultivations.

The plough is now ready to take to the field. Next it must be hitched to the tractor correctly and the whole outfit lined up.

Reversibles

In order to get well matched ploughing in both directions from a reversible plough, all measurements and settings on the left-hand set of bodies must be the same as those on the right-hand bodies.

Working on a level surface, carry out the following checks.

If pitch varies from share to share, uneven work is inevitable. Fit a new set of shares, then as a rough check, place the straight edge along the line of the shares to see it touches all of them. A more accurate method is to measure vertically from the share point to the underside of the beam (Fig. 6.9). To adjust, place shims under the share. On some ploughs pitch can be altered by an eccentric bolt.

If mouldboards are not the same distance apart all along their length, uneven width furrows will result. If they are too wide, draught will be excessive and the furrow will roll back. Check them by measuring the distance between

Fig. 6.9

share points and then the distance between the backs of the mouldboards.

The mouldboards can be adjusted by the tie rod behind each furrow, or again, by shims. Final setting can be done in the field.

Lateral distortion of the legs can be measured by placing the straight edge at right angles to the main frame near the share points, and using the tape to check that the offset between point and straight edge is the same on both sets of bodies.

The rear landsides must be parallel and in

Fig. 6.10

line with each other. Swing the plough on its side, place the straight edge across the main beam and at right angles to it, and check that both landsides are an equal distance from the straight edge (Fig. 6.10).

If the turnover mechanism is hydraulically operated, check the couplings and hoses for leaks, cracks or perishing. If it is mechanical, then chances are it will have worn with use and the plough will turn over with a bang. This may distort the plough, so adjust the linkage to give a nice easy turnover.

Before attaching the tractor to the plough, check front and rear wheel settings in the maker's handbook. Failure to do this will cause the tractor to crab, resulting in an incorrectly sized front furrow, and will place extra strain on both tractor and plough. In heavy conditions it may be necessary to add weights and water ballast to the rear wheels. Make sure each wheel is equally ballasted. On hilly land, or when lifting out of work, weights must be fitted to the front of the tractor, because a reversible plough is so heavy that the front wheels will lift off the ground and the tractor could overturn.

Also, the lift rods must be the same length or else the plough will tilt (Fig. 6.11).

Fig. 6.11

In-the-field check

Having checked over the plough in the workshop and attached it to the tractor, we must do the remaining settings in the field.

To get well-matched bouts from both sets of bodies, various adjustments, including carefully setting levels across and along the plough's axis and front furrow width, are essential. Failure to adjust level settings properly will give a different depth of ploughing on one set of bodies. A small front furrow width means it will take longer than necessary to plough the field, and give an uneven-looking finish to the furrows. Wide front furrows will increase draught and if too wide the furrow will stand on its side and not bury the surface rubbish. A depth to width ratio of 1:1½ is a good average.

Plough manufacturers have various ways of altering furrow width and tilt, so the handbook should be consulted at all times.

Top link setting will affect lengthways levelling of the plough. Too short a top link gives a deep front furrow with the last one much shallower and broken furrow wall (Fig. 6.12) and the converse for too long a setting on the top link. Check for even depth of furrows with a tape measure just behind the mouldboards (Fig. 6.13). Another way is to squat a few metres away from the plough's side and by eye check that the beam is the same height above ground at both the front and rear furrows. A correctly-set top link will allow the rear landside to just press on the bottom of furrow and keep a clean furrow bottom (Fig. 6.14).

Fig. 6.13

Fig. 6.14

Headlands should be carefully marked out in order to leave sufficient space for the plough to be lifted out of work on the move. Many ploughmen stop when the front wheels touch the hedge and lift the plough out of solid ground, overloading the hydraulics on the tractor. Remember also when ploughing out the headlands to use both sets of bodies equally, otherwise one will wear more than the other. Skim and disc coulter settings are as for conventional ploughs.

Fig. 6.12

One final point, if the plough is to be left idle for any length of time — paint the shiny parts with fresh oil to prevent rust and subsequent scouring problems when ploughing starts again.

Hitch, draught and wheel setting

The plough must be hitched up correctly to make a workable unit. If the top link is in line with the centre line of the tractor the draught line is correct (Fig. 6.15). Crabbing increases drawbar pull and every extra 90 kgf demands an additional one kilowatt.

Fig. 6.15 When ploughing the top link and plough frame should be in line with the tractor.

The power needed to pull a plough is calculated by the following formula: the furrow width in metres x depth of furrow in metres x number of furrows x soil resistance.

For example, a 4 furrow plough with 350 mm bodies working 250 mm deep in medium to heavy loam would require a drawbar pull

of: 0.350 m x 0.250 m x 4 x 70000 N/m^2 = 24500 newtons drawbar pull. Remember that tractor power is measured in kilowatts and 1 watt = 1 Nm/s (see page 8). So if you know the speed of ploughing, you can calculate the drawbar power in kW required to pull the plough. For example, if the plough discussed above were to be pulled at 10 kmph it would require

$$24\,500 \times \frac{10 \times 1{,}000}{60 \times 60} \text{ watts} = 68 \text{ kW.}$$

drawbar pull x speed in metres per second

Soil resistance varies from 35000 N/m^2 in light sandy soils to 100,500 N/m^2 in heavy clay. An average of 70000 N/m^2 would apply to medium to heavy loam. Always take the farm's heaviest soil in making calculations.

If rear wheel setting is wrong there will be difficulty in getting the correct front furrow width. This may be obtained by changing the position of the cross shaft in relation to the main frame or by turning the cross shaft adjusting lever. But this does not ensure a true draught line. So before coupling the plough to the tractor see that the rear wheel centres are correct according to tyre size and front furrow width. On 12.4—36 (11 x 36) tyres the tractor rear wheels need to be at 1320 mm centres for 250 mm wide furrows, 1420 mm centres for 300 mm furrows and 1520 mm centres for 350 mm furrows. On 16.9—30 (14 x 30) tyres, 1420 mm centres are needed for 250 mm furrows, 1520 mm centres for 300 mm furrows and 1625 mm centres for 350 mm furrows.

Altering rear wheel centres may necessitate moving the front ones. The inside of the front tyres should be in line with the inside of the rear for ploughing.

For coupling on uneven ground first attach the left-hand link, then the right-hand one, using the linkage levelling and cross-shaft adjusters to line up the pin, and finally the top link. Check chains should be slack when the plough is in work but tight enough to prevent the plough from swinging and fouling the tractor tyres when it is lifted.

Hard-surfacing ploughshares

The rate of wear on the cutting edges and points of a ploughshare can be slowed down by hard-surfacing them. This way share life can be doubled.

Tools required are an oxyacetylene gas welder, a portable grinder or file, rods that deposit abrasion-resistant metals and a basic knowledge of gas welding.

Hard-surfacing the new share shown here took one rod. A partially-worn share is more expensive to do, requiring more rods and time to build it up to its original level. Also, soil will have been driven into the metal, making it difficult to clean. Impurities will float to the surface of the weld pool, interfering with the fusion process and giving a poor weld (Fig. 6.16).

Fig. 6.18 The new share should be cleaned down to the bare metal.

Fig. 6.16 A new share (left) takes less time to hard-surface and requires less deposit.

Fig. 6.17 Areas to be treated are shown by the chalk-marks.

First, areas to be treated should be cleaned down to the bare metal (Figs. 6.17 and 6.18).

The rod used in the job illustrated was a wear-resistant alloy steel that deposited small amounts of carbon, silicon, manganese and other metals in the weld.

Ordinary rods will not do, as they are generally intended to give a soft deposit that is strong and ductile. Alloys that resist abrasion have higher degrees of hardness and are prone to crack under impact. Check with your supplier that the rod will do the job.

A No. 10 nozzle was used with pressures of 0.25 bar on both oxygen and acetylene bottles. Flame length is about two-and-a-half times the ordinary neutral cone length. This will help deposit more carbon in the weld.

Hold the rod and torch at an angle of about 25° to the work. Pre-heat a small area of share and as soon as the metal starts to 'sweat' and bubbles begin to appear on the surface dip the rod into the sweat pool and melt the rod on to the share. Take care not to get the parent metal of the share too molten or there will be too much mixing of the rod deposit and the parent metal, giving a softer deposit.

Make a bead about 12 mm wide on the

top surface of the share, using a weaving, up and down motion with the rod. Underneath the share point hard-surface an area from the point to 25 mm farther back. Lay a narrow

Fig. 6.19 Laying a bead along the front cutting edge.

Fig. 6.20 Discs must be renewed when they get into this state.

bead along the land side or 'suck' edge (Fig. 6.19).

Finally leave to cool and wire-brush off any scale or slag.

Much cheaper ways of hard-facing have now been developed and these involve less skill. Usually they do not create such thick layers of deposit but are otherwise just as satisfactory. Spray welding with a special torch or hard-facing with a paste and conventional torch are useful techniques for the farm workshop. The equipment and techniques are described on pp. 160—162.

Hot riveting a disc coulter *(Fig. 6.20)*

To renew disc coulters first place the disc in a vice and grind or chisel off the rivet heads (Fig. 6.21). Do this from the disc end of the rivet, not from the casting end, otherwise you may crack the casting with the chisel or bite into it with the grinder.

Carefully punch out the old rivets and clean off dirt on the flange to ensure proper seating of the new disc (Fig. 6.22).

Use the gas welding torch to bring the rivet to forging heat, then push the rivet through the casting, place the head on a solid surface —

Fig. 6.21 Remove rivet heads from the disc side to avoid damaging the casting.

a blacksmith's anvil is ideal — put the disc over the rivet and seat it on the casting flange.

While an assistant holds the disc to keep the rivet head on the anvil, tap the rivet with a hammer to spread it in the hole. Use a snap to form the head. Repeat process with the other two rivets. Figure 6.23 shows the finished job.

Fig. 6.22 Punch out the rivets through the back of the casting. Take care — a mis-hit may crack the casting.

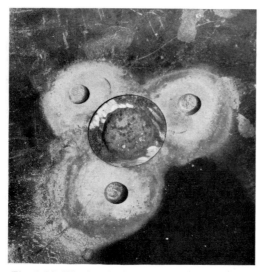

Fig. 6.23 The finished job — a tight joint with three nicely rounded rivet heads.

ROTARY OVERHAUL

There are probably 30000 rotary cultivators on British farms waiting for the soil to dry out. Many will have been dumped in the back of the implement shed just as they were when the weather stopped further autumn work. They are hard working machines built to take a severe battering, but they still need maintenance.

This text is based on a Howard E series 60 in Rotavator and most hints apply to the rest of the range. For many other models the principles apply but the instruction book should have the final say on details.

Mount the machine on the tractor linkage ensuring first that the mounting plates holding the hitching points are set right for the particular tractor. And, as there are four lower link hitch positions, mountings 1 and 3 or 2 and 4 should be used respectively for either offset or in-line working. Do not splay the links to combine positions 1 and 4.

Detailed settings will be given later. Levelling the machines is the first aim. The rotor and the tractor axle are the guide lines one way; the gearbox is used to decide the front-to-rear level. Adjust the tractor's offside lower link and the top link accordingly.

Use the grease gun with a lithium based grease. Apart from nipples in the universal couplings at both ends of the pto drive shaft there are others on the depth wheel, at the pivot point of the shaft holding it and at the offside rotor bearing (Fig. 6.24). These should be daily jobs during work. Pump in grease until the new stuff shows — except in the rotor bearings. This will take about a dozen pumps of an average gun to fill it then three or four pumps a day.

It is worth pulling apart the two sections of the drive shaft to clean and smear with graphite or molybdenum disulphide grease.

Next check for tightness the castellated nut

Fig. 6.24 Rotavator grease points.

The depth wheel greasing point. The scraper can be adjusted by nuts (A).

Grease the depth wheel arm pivot here. (B) is the drive chain tensioner.

Fig. 6.25 The castellated nut here must be tight . . . (see Fig. 6.26).

Grease nipple at the offside end of the rotor. (C) is where the blade straightening bar stows. (D) adjusts the skid height.

Fig. 6.26 . . . otherwise the rubber sealing ring (indicated here) will allow the gearbox oil to drain out.

rather awkwardly situated to hold the drive shaft on the gearbox — particularly if this was not factory fitted when the machine arrived (Figs. 6.25 and 6.26). If it has worked loose the chances are that oil will have leaked past the rubber sealing ring inside and will continue to do so when you have drained, flushed out and refilled the gearbox, which is the next job. Avoid any chance of the gearbox oil draining out of the breather hole by parking the machine properly.

At the same time check the Selectatilth gears. There should be four under the cover, two driving and two 'spares' for changing rotor speeds which fit against the working pair and keep them up to their job. When replacing these gears, beware of putting them in the wrong way. One side has a slightly protruding face which should always face forwards. It would work the other way but metal chippings would be ground off into the gearbox (Figs. 6.27 and 6.28).

Fig. 6.27 Check position of Selectatilth gears and the spares in the cover. (E) is gearbox oil level.

Fig. 6.28 The raised edge of the gears must face the front of the machine.

The encased chain on the nearside which drives the rotor from the top shaft needs attention next. Drain, flush and refill — gearbox and chaincase both take SAE 90 on this machine — and if the case has not been off for 500 hours, remove it and clean it with paraffin.

Chain tension needs checking more often, however, and with the casing on, this can be done through the inspection plug on the rear side. A piece of wire and a screwdriver pushed against the chain inside will soon show if the correct 20 mm amount of play takes place.

Adjustment can be made on the screw with its locknut on the leading edge of the case (Fig. 6.29).

Fig. 6.29 Check for 20 mm play on drive chain by poking with a screwdriver. Turn bolt (F) to adjust.

While the toolbox is this side of the machine, look at the scraper on the depth wheel and adjust the clearance over the wheel by the two bolts in their slotted holes.

The Rotavator must be properly shod. The right number of blades should be fitted to each flange and each one should be in reasonable condition. If a two bladed configuration was used to produce a rough winter seed-bed it may be changed to a three-bladed set-up for finer spring work. Alternatively, it is simpler, cheaper and faster to leave the 2 blade rotor and speed it up as necessary with the implement's gear box.

The makers advise changing all blades at once; 'bumping' will occur in work if some are better than others. So long as a blade has not worn to a point there is still some life in it.

Three main points to remember during reblading are that the left-hand blade of a pair leads every time, that the 'scroll' pattern across the rotor must be maintained to give the follow-my-leader penetration evenly from side to side, and the blades must be tight (Fig. 6.30).

Howard's freely replace broken, half-worn blades, but not if they have broken through the bolt holes — a sure sign of slack nuts. Use only the proper nuts and bolts (Fig. 6.31) —

Fig. 6.30 Left-hand blades must lead each pair. This right-handed one is worn but the mark shows when it is time to change.

Fig. 6.31 Use the right nuts and bolts for the job and keep them tight.

Fig. 6.32 (a) If the shank is bent straighten it with the bar provided. Otherwise uneven running, extra wear and increased power requirement will result.

the nuts are long and fine-threaded like car wheel nuts. Put nut and bolt on the right

(b) The blade straightening bar can also be used to help move the machine for hitching-up.

way round, with washer and nut against the rotor flange and the bolt head on the blade side. If blade fitting is a straight replacement job remove one and fit a new one, rather than strip a flange and have to plot the pattern again.

The shanks of some blades may be bent. A special tool simplifies straightening them (Fig. 6.32).

The shield is not just a safety cover. Its position plays a vital part in the state of the final tilth. Oil the hinges and make sure the springloaded slot which holds the check chain is in good order. Hold the shield in place with the top clip to reduce shock loads during transport (Fig. 6.33).

Fig. 6.33 For transport, hook the cover right up, using the hook on the cover itself. Height for work is set using the chain in the spring-loaded slot (A).

On the flange at each end of the rotor are the weed shearing blades designed to prevent rubbish wrapping round. They fit any pair of bolt holes to suit the blade formation. The fixed shearing plates, are slotted and two set-screws each side need slackening to allow the plates to be adjusted to just clear the cutters (Fig. 6.34).

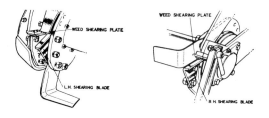

Recommended order of hitching is top link, nearside lower link and then, using the blade straightening bar as a lever, the offside lower link can usually be kicked into position. For safety have the pto out of gear and the tractor switched off before coupling up the shaft.

Fit the stabiliser bars on the machine's mounting pins, raise the linkage and fix the other ends on the tractor. If chains are used tighten to limit side sway to 25 mm.

The top link should be adjusted so that with the blades on the ground the gearbox is horizontal. Level off across the machine by winding down the depth wheel until the blades are just clear, then wind the tractor's offside lower link adjuster so that blade clearance is equal right across.

Fig. 6.35 The four mounting points can be seen.

Figure 6.35 shows the mounting points. These plates can be bolted in a variety of positions to accommodate the varying length of tractor lower links. It is essential to find out from the instruction book which position the plates should be in before the machine is changed from one model of tractor to another.

The skid on the opposite side to the depth wheel should be set to run about 25 mm above the finished work. It is not a depth control skid and is simply there to prevent the blades digging in too far if the depth wheel drops in a hole on uneven ground. Before starting work it is vital that the slip clutch is set exactly as detailed in the instruction book.

To rub home the message of the previous pages on overhaul, the following photos show some expensive reminders of what can happen when maintenance is slack or the machine is misused. The same principle applies to many machines on the farm (see Fig. 6.36).

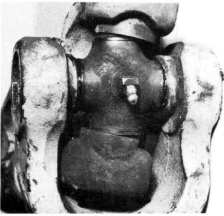

Fig. 6.36 Damage caused by improper maintenance.

The universal joint collapsed through lack of proper lubrication. The left and right bearings were greased but it was not forced through the top and bottom ones which have broken down.

This damaged crown wheel was caused by faulty clutch setting. The tension nuts should have been tightened until each spring was compressed 4 mm measured between the sheet metal cover and the flat washer to give a breakaway torque of 1356 Nm.

This damaged gearbox was caused when the top stays were allowed to work loose.

This back plate was twisted when the machine was turned in work without lifting it out.

Lack of lubrication has caused the needle rollers to disintegrate and the clutch drive plate yoke took the full force of the misalignment.

HARROW CARE

Harrows must rate among the toughest and often the most ill-used pieces of equipment on the farm. Fortunately they are usually built to take it, but care in use and treatment can mean better work and a longer life.

There are scores of types but, generally, those with shorter 60 mm tines should be used on grass and the 110 mm tined units kept for arable. Unless there is a very thick matt of grass the longer tines will tend to jump about and do little good. Best speed is 6 to 7 kmph to let the harrow do its work rather than the highest possible gear at full throttle.

Take greater care using a new harrow. The sharper tines will bite more than a worn set. Pull away gently — jerky starts soon result in something breaking.

Harrows should not be loaded with sleepers or other weights to make the harrowing effect more severe, nor should they be pulled sideways through a gateway. Both actions cause excessive loads on the links.

If a link breaks, repair it at once with links of the correct size — otherwise more strain will be put on the surrounding links, setting up a chain reaction of breakages. If the job needs two harrows in tandem take the towing chain of the rear harrow over the front one and attach it direct to the tractor — don't just hitch it to the harrow in front.

Check harrow drawbars or whippletrees for alignment. If they are bent, more strain is placed on a few pulling points. Careless driving through gateways or near banks, posts, trees and other solid objects is the main cause of bent drawbars.

Hydraulic lift mounted harrows can serve two purposes. A heavy harrow can be made lighter by slightly raising the linkage and, on dirty ground, flexible harrows are much easier to clean.

If repairs become necessary remember that flexible harrows are usually made of high carbon steel and cannot be welded. To fit new links, heat the link where required and fit it to the harrow. Do not cool it using water or it will become brittle and fracture. Let it cool naturally.

DISC HARROWS

Disc harrows are among the hardest-working implements on the farm, but properly maintained they will last for years. Routine maintenance in the field should include daily greasing of all the bearings and the pivot points that angle the discs. Try to make sure no dirt stays in the bearings by pumping grease in until you see clean grease coming out. Dirt and grease form an abrasive grinding paste that quickly

Fig. 6.37 Set the scrapers close to the discs.

Fig. 6.38 Tighten discs with stillsons and a spanner.

wears out bearings.

Main field adjustments are the top link and tractor lift rod setting for levelling the equipment, and scraper settings. Set scrapers as close to the discs as possible to avoid soft build-up in sticky conditions. Knock bent scrapers back into position with a hammer (Fig. 6.37). If they are badly bent and cannot be straightened with a hammer, take them off the machine, heat the bent portion with a cutting torch and straighten them on the anvil.

Each gang of discs consists of a square centre shaft with each disc separated from its neighbour by a metal spool. Spools and discs are kept tight against one another by a thread and nut on one end of the centre shaft. During work, wear on spool faces and disc centres will cause the discs to become a sloppy fit on the shaft and to wobble. Tighten the nut on the end of the shaft of each gang before this becomes too serious (Fig. 6.38). Use thick flat washers or spacers to take up excess wear at the end of the shaft, otherwise the discs will eventually revolve independently of the shaft instead of with it. The net result will be expensive and unnecessary replacement of discs and shaft. Tightening the discs is a simple job. Hold the end spool nearest the adjusting nut with a pair of stillsons to prevent the shaft revolving and use a ring spanner on the nut.

Partly-worn disc centre holes may be reduced to their original size and shape by laying them on an anvil, concave side up, and peining with a hammer round the edge of the hole. Clean up the rough edges.

Discs bearings take downward loads from the discs and frame and side or radial loads from the curved discs. Most common types of

bearings are chilled cast-iron centres and caps, with flanges on the centres to counteract side loads. Two bolts hold the bearing together. Another type is a cast-iron centre with replaceable, oil impregnated hardwood shells in the caps (Figs 6.39 and 6.40).

Fig 6.39 Replace cast-iron bearing centres and caps.

Fig. 6.40 Wooden bearing shells.

With all-cast-iron bearings renew both the centre and the caps. Renewing just the centre is false economy, as wear also occurs on the caps. Putting a new centre in old caps will allow the centre to flop about and wear out rapidly. Wooden shells are cheaper, but will need replacing more often. Make sure all dirt is removed from the caps, otherwise new shells will not seat properly and protrude above the edges of the caps. When you tighten the holding bolts the shells will split.

Other points to check include all hardware on the implement, cracked or chipped discs and wear on the bumper or end plates on each gang, particularly where the ends of the front gangs rub together.

MANURE SPREADER OVERHAUL

The manure spreader, above all other farm tackle, works under severe conditions and requires extra care.

A high pressure water hose, grease-gun and oil-can rank as the spreader's best friends. Regular hosing flushes off the ammonia and corrosive salts in the muck and prevents them attacking the metal and paintwork.

Worn and broken parts can then be spotted before they cause a major failure, grease nipples are in sight and overhauls are easier.

When doing an overhaul take each section

Fig. 6.41 Important maintenance points on a ground driven machine.

If wheel end float is more than 2 mm the contact area of the drive pawls is reduced.

A broken pawl spring (see A) will reduce the effectiveness of each feeding stroke. The split pin securing the ratchet wheel (see B) is incorrectly fitted.

This is the kind of fault that shows up once the muck has been washed off.

and drive of the machine and follow it right through. The best point to start is on the drive mechanism.

Ground driven machines have many more working parts and require more attention (Fig. 6.41).

First check the pawl and ratchet drives in both wheels. Remove each wheel in turn and wash off the old grease and dirt. Inspect for wear, especially the pawls and springs. Leading corners of the pawls become rounded and should be renewed or built up with weld. Weakened springs may be temporarily strengthened by fitting small washers or spacing pieces in their locating holes, but take care not to make them coil-bound.

One wheel hub will carry the main drive sprocket. The chain must drop centrally on to the sprocket and be fully engaged when the lever is in the drive position. The sprocket has to withstand the full strain of all the driving force required and if the chain is incorrectly adjusted it will jump over the teeth.

The opposite wheel operates the conveyor chain feed mechanism, usually consisting of a cam with three or four lobes, a ratchet wheel and actuating lever. Check on the amount of wear in ratchet pawls, condition of the springs, lever pin wear and the efficiency of the linkage return springs.

Ensure the tyres are correctly fitted. The point of the 'V' tread bars should be trailing in direction of rotation — pointing to the rear of the spreader. Correct air pressure is important to the performance of ground driven machines. They rely on the grip obtained by the tyre for all the drives and a too low pressure will result in poor spreading and tyre wall damage as well as increased draught.

The wheels should have no more than 2 mm end float to ensure the correct engagement of the pawls inside the hubs. Special stepped collars are usually fitted to enable any excessive free play to be taken up. If there is no adjustment left on the collars, a washer fitted between the collar and the hub will do the job.

Drive shafts

On power driven machines, where the main drive shaft runs under the body, it is worthwhile to check that the shaft has not been damaged by driving over rough tracks or through rutted gateways.

Some pto models have square section drive shafts and these can be coupled up incorrectly. If the universal joints are not in line, as shown in Fig. 6.42, jerky drive will result and play havoc with the needle roller bearings.

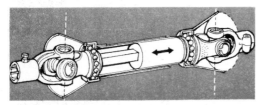

Fig. 6.42 Pins, represented by dotted lines, must be assembled in parallel.

Bevel gear boxes require little attention apart from keeping an eye on the oil level. Overfilling can lead to a lot of trouble with burst oil seals. A gearbox with too much oil builds up a pressure within the housing, caused by the pumping action of the gears. Some gearboxes have special breathers to allow the pressure to escape.

Safety clutches incorporated in the pto drive shaft should be dismantled, cleaned and re-assembled dry. For shafts fitted with shear pins, rather than clutches, it is best to remove the pin and check that the bushing has not seized. This type of safety device may be oiled to ensure it will be free should the pin shear.

Sliding couplings become dry and gummy and should be cleaned off regularly and greased to prevent excessive end loading in the joints.

Open-link chains

Open-link chains, widely used on spreaders, must be correctly fitted, especially where the driven sprocket is smaller than the driving sprocket. The rule is: run drive chains with

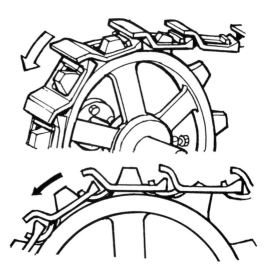

Fig. 6.43 (a) Hooks forward and slots to the outside on drive chains.

(b) Bars forward and slots outside on conveyor chains.

hooks forward and slots to the outside, see Fig. 6.43(a). For conveyor chains the bars should be forward and slots outside (Fig. 6.43(b)).

Correct chain adjustment is vital. If they are too tight they will cut into the sprocket teeth and put excessive loading through bearings and shafts. If they are too slack they may ride on top of the teeth and whip excessively.

If the links can just be moved from side to side across the sprocket teeth, the tension is about right. If 33 links of any open link chain have worn so much they are one link longer than 33 new links, replace the chain.

Feed conveyor chains must be adjusted to equal tension on both sides or they will run off-centre and cut into the body of the speader. See that slats are running at right angles to the side boards and are correctly meshed with the drive and idler sprockets. Chain sag should be about 3 per cent of the floor length.

Lubricate them at frequent intervals with a light engine oil. An oil paint-brush makes an ideal applicator, and to ensure the oil runs into the joints it should be applied to the open side of the links. Roller chains are best oiled on the side that contacts the sprocket teeth. Note the incorrect position of the tensioning sprocket in Fig. 6.44.

Fig. 6.44 Chain faults. The tensioning sprocket is fitted to the wrong side of the chain thus reducing the number of links in contact with the driving sprocket.

The quality and evenness of spread is governed to a great extent by the efficiency of the shredding cylinders and distributor paddles. If tines are missing or paddles broken off the remaining parts come under increased strain. This kind of overloading brings about chain and sprocket wear, as well as bearing wear. It also causes lumpy spreading (see Fig. 6.45).

Cylinder shaft bearings should be checked for wear, thoroughly cleaned and greased.

Baler twine can damage shafts and bearings, causing overheating in the bearings and melting the grease. Sealed bearings are most vulnerable, the twine wearing away the sealing washers and leaving them open to liquid and solid manure.

Always start loading at the opposite end to the shredders and fill until the lower cylinder is reached. On no account pile manure on top or

Fig. 6.45 Maintenance points of the shredding and distributor mechanisms.

Check that shaft securing collars have not worked loose.

The top shredder cylinder on this machine will do little to help even spreading. The majority of tines are either bent, broken or missing altogether.

Bent and twisted distributor paddles reduce the effective spreading width.

force it in and around the shredders. Leaving them clear allows easier starting and reduces the initial strain on the driving mechanism. Loading in this manner gives a more even feed because the manure is more easily separated.

Make it a habit to engage the spreading drive lever before the feed lever. This gives the shredder cylinders time to reach their correct working speed and clear themselves before the main bulk of the load starts moving.

In cold weather, make sure that the conveyor chain is not frozen to the bottom of the spreader before loading.

End-of-season maintenance should be done as soon as the last load has been spread and before the muck cakes and hardens. Remember to hose off the parts underneath the body.

Chains should be removed, washed in paraffin and given a coat of heavy grease before being refitted.

Grease the conveyor chain and give spreading gear and the parts where paint has been removed a liberal coating of creosote or old sump oil.

THE ROTARY MANURE SPREADER

The rotary type of spreader has the advantage of simplicity and low maintenance combined with the ability to handle virtually all kinds of manure. A fairly high powered tractor is required to operate them but alternative drive sprockets are usually available to allow the machine to be driven by a less powerful tractor.

One of the key operational points is to wind the chain around the rotor before the drum is filled — this is easily achieved if the pto is engaged very slowly when the drum is about half full and the rotor allowed to revolve about ten times.

Remember not to pack the ends too tightly when loading or more power will be required in order to start the rotor turning.

Due to the high torque on the pto when spreading, nearly all manufacturers state that the pto should be disengaged before turning. Failure to do so could result in damage to both the machine and tractor.

The most likely causes of failure stem from broken or stretched flail chains but this kind of damage can be averted by periodically cleaning the machine and giving it a quick inspection. At which time the following points should be checked.

(1) Remove any string which may have wrapped itself around the rotor (see Fig. 6.46). Do not burn it off as the rotor bearings will almost certainly be damaged.

Fig. 6.46 Remove any string which is wrapped round the rotor.

Fig. 6.47 Check the rigid flails.

Fig. 6.48 Shorten stretched chains by inserting bolts.

(2) Check that the two rigid flails at each end of the rotor are not missing or seized up (Fig. 6.47). It is essential that they swing out easily to help start the chain's spreading action.

(3) Make sure that the chains have not stretched to such an extent that the flails are beginning to strike the inside of the drum. Fitting a bolt at the inner end of a chain will shorten it (Fig. 6.48).
(4) The rotor bearings on the outside of the spreader need regular greasing and inspection (Fig. 6.49). These are very prone to damage and it is not uncommon to find that the seals have broken open to expose the bearings to dust and dirt. Fit new bearings as soon as possible. Failure to do so may result in the bearings collapsing and allowing the shaft to drop which leads to .the flails hitting the drum and the drive chain jumping its sprockets.

Fig. 6.49 Inspect and grease the rotor bearings.

Fig. 6.50 The main drive chain should have 12 to 18 mm deflection.

(5) The main drive chain should be maintained at the correct tension which is about 12 mm deflection (Fig. 6.50).

(6) To adjust chain tension slacken off the four bolts which retain the lower sprocket bearing housing and move the housing down using one of the adjusting nuts. It is advisable to remove the chain periodically and wash it in paraffin. Soak it in clean engine oil with a molybdenum disulphide or a similar additive for a couple of hours before re-fitting (Fig. 6.51).

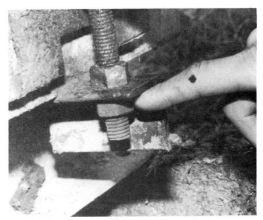
Fig. 6.51 Main drive chain adjuster.

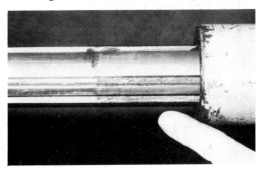
Fig. 6.52 Lubricate sliding section of the pto shaft.

(7) A rotaspreader requires a high torque on its driveshaft so it is important to the shaft to telescope easily. It should be regularly separated, cleaned and lubricated. Failure to do this can lead to the universal joints being damaged. These joints should also be greased daily (Fig. 6.52).
(8) Do not forget to make routine checks. The wheels themselves should be inspected for severe corrosion (Fig. 6.53). It is quite common for a tyre to blow off the wheel due to one wheel rim collapsing because of corrosion.

Remember — missing chains or flails unbalance the rotor and cause damage to its bearings by vibration. Check the condition of the bolt holes of the lugs which secure the chains (Fig. 6.54). A chain flying off during work could cause an accident.

Fig. 6.53 Check the condition of the wheels.

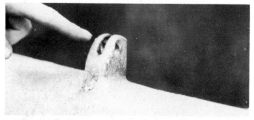

Fig. 6.54 Inspect the flail chains' mounting lugs.

MOWER OVERHAUL

In spite of the various alternatives now available to the finger bar type mower, this is still very widely used and is likely to require more attention than other kinds.

Though different makes of finger bar mower may vary slightly they usually have the same type of cutter bar assembly and method of adjustment and alignment.

The first job is to strip down the entire cutter bar and thoroughly clean this and all parts before checking fingers, wearing plates, ledger plates and clips for wear or damage. The knife will also need to be checked and new sections fitted where necessary, others being sharpened (see illustrations in Fig. 6.55).

Having thoroughly overhauled the cutter bar, the knife is replaced and properly bedded down and the cutter bar aligned. Failure to attend to these points will reduce efficiency of the machine and undo much of the work that went into the overhaul. The mower should be

Fig. 6.55 Cutter bar overhaul.

(a) Completely strip the cutter bar, arranging the parts in order after removing all dirt, etc., with a wire brush. Check bar for signs of damage. The points of the fingers should be rounded, not pointed, preferably on a bench grinder. The sides of the fingers and liners are also ground to give a sharp edge to the liners. The aim is a smooth pointed finger and sharp shearing edge for the knife.

(b) (Left) The angle at each side of the finger should be ground to leave an angle as indicated by the position of the screwdriver. On some makes the finger liner can be replaced and where wear is apparent this may postpone fitting a new finger.
(Right) Wear on the clips should be checked. A new clip (top) is compared with a worn one. This needs replacing because the face has become badly worn, either through wear or damage.

fitted to the tractor during this work.

The pitman must run at right angles to the crankshaft and the knife should operate when in work in a straight line with the pitman. The

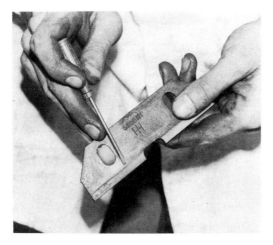

(c) Wearing plates have a slight recess at the leading edge. Check that this is not deepened or extended through wear or poor adjustment when a new part will be necessary.

(e) Rivets should be neatly tightened and rounded with a proper punch. This gives a more permanent job than where they are tightened with a hammer. The knife sections must always point away from the worker.

(d) Worn knife sections should be removed by placing the knife back vertically in a vice and giving the rear of the section a sharp blow with a hammer. The rivets are punched out as shown. The rivets on the section nearest the camera have been badly flattened with a hammer and will soon loosen.

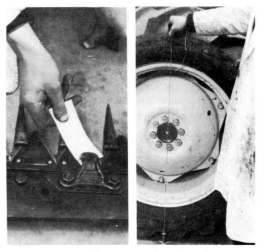

Fig. 6.56 Setting up the cutter bar.

angle can be checked with a straight edge and altered by adjusting the brace and yoke bars. Knife angle can also be corrected by adjusting the mower headstock and, since the fittings vary, it is necessary to check the manufacturer's instruction book for method of adjustment.

Bedding down the knife and alignment of the cutter bar are illustrated in Fig. 6.56. When this has been completed the mower should be greased and the gearbox oil level and drive belts (where fitted) checked.

(a) (Left) Check tightness of clips with a post-card. If they are too tight, they can easily be corrected by placing across a vice and lightly tapping with a hammer. On some models shims can be used to adjust the clips. Wearing plates can be adjusted to run level with the knife. Check each one and adjust to correct position so that there is no 'fore and aft' knife play.

(Right) When the knife has been bedded down, check alignment. First, drop a plumb line across the centre of the rear tractor axle and mark the base position on each side.

(b) The first job is to get the knife to run smoothly without any 'crabbing'. Fingers which are not level can be tapped into position or a tool, consisting of a piece of tube welded to a bar, can be made (as shown here) to ease the fingers into line. Check that fingers are in line by pulling a piece of string across them.

(d) Always press the end of the cutter bar back before measuring to take up any play. The mower is then raised on the tractor linkage and the height of the cutter bar measured under each end. The cutter bar should leave the ground horizontally for the first few centimetres and thereafter the outer end should gain on the inner end so that it is well angled in the raised position. Correction is made by adjusting the tension springs.

(c) The tractor is then moved forward, a straight edge put across the two marks and a line drawn in front of the cutter bar. Measuring from the rear edge of the cutter bar to the line at each end will establish the 'lead' which can be adjusted by tightening or slackening the braces or yoke at the headstock. The amount of lead should be 20 mm in 1 m of cut so that in work the cutter bar runs in a straight line with the pitman.

will butt together. Because of this it is vital to lay sufficient weld metal for the join to be hammered out on an anvil so that the new hole can be drilled in the exact place of the original.

The choice of welding rods is between mild steel, silicon manganese or 3% nickel steel. Mild steel is easy to weld but lacks high tensile strength and is likely to break. Silicon manganese is a more difficult rod to use, but

After the cutting season the knife should be removed and stored with the spare knives. The cutter bar, knives and all working parts should then be covered with grease or a rust preventative compound.

The pitman is designed to protect the knife and drive from damage, but there are occasions when the knife back is broken by stresses in work. A spare knife is usually available, but if replacements are limited it is a reasonably simple matter to weld the broken pieces to avoid renewal and reblading (see pictures in Fig. 6.57).

Usually the knife back breaks across a hole near the drive end where stress is greatest. The two broken faces must be ground down so they

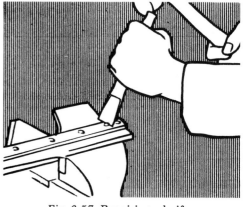

Fig. 6.57 Repairing a knife.

(a) First remove the broken back from the knife head and also two or three sections next to the break on the other piece. If the break occurs in the centre of the knife simply remove a few sections on either side of the break.

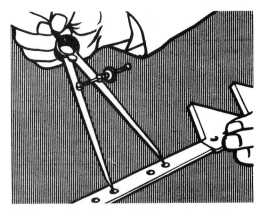

(b) Measure the distance between three holes with dividers to ensure proper spacing.

Do this before welding or heat expansion may cause errors. File broken parts level. Make sure the knife back lies straight when the filed edges are placed together. Weld with a good build-up on both sides (right), preferably using a No. 5 nozzle with a 2.5 mm diam. rod and neutral flame.

(c) The oxygen cylinder pressure should be 0.25 bar and the acetylene 0.125 bar. While the weld is still hot, place the knife back on an anvil and hammer out until the weld has been expanded sufficiently to allow a fresh hole to be drilled in the right place.

the work produced has an extremely high tensile strength, as it has 3% nickel steel. The latter has a high heat tractability so care must be taken in cooling the work.

Silicon manganese is probably the best choice for the farm welder. Properly used, it will result in a lasting repair in which the weld will be stronger than the parent metal. The illustrations in Fig. 6.57 show the repair method using oxyacetylene equipment.

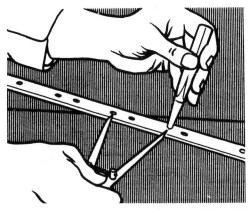

(d) Cool the work slowly before checking with dividers to ensure proper spacing then file off excess weld metal and centre the position for the hole with a punch.

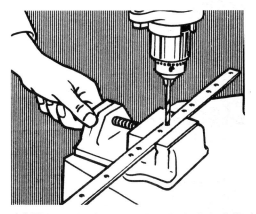

(e) This method ensures that the hole is drilled through the centre of the weld to give maximum strength. Drilling on the edge of the repair may weaken the weld.

ROTARY MOWERS

Getting the best from a drum mower depends on two things: careful, skilled driving and good maintenance. These factors are examined here and applied, as an example, to a four drum mower, the Bamford C450.

The basic objective of a mower is to cut while in the working position. So driving should be directed at just that. Run a mower at rated pto speed but only when in working position. Make sure the safety guards are in position. (Even if you don't employ labour, remember that the safety regulations are based entirely on accident statistics.)

The parts that will wear out most rapidly are the cutting blades. Changing a full set takes five minutes or so. Use the knife tool provided to

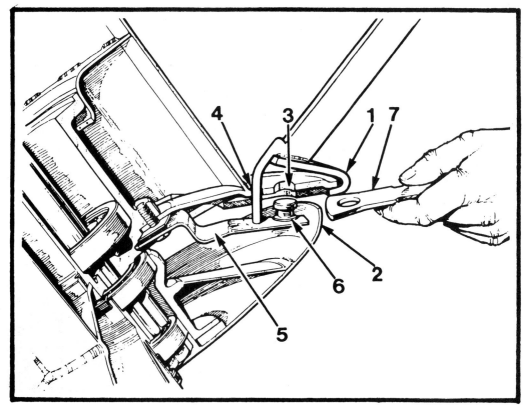

Fig. 6.58 Rotary drum mower — blade replacement.
Key: 1 Knife tool, 2 Saucer, 3 Hole for knife retaining stud, 4 Hole for knife tool, 5 Knife retainer spring, 6 Knife retaining stud, 7 Knife.

depress the knife retainers. Don't over-do this as it may damage the spring-steel of the retainer. New knives should be fitted with the chamfered edge of the knife upwards (Fig. 6.58).

The saucer under each drum is fitted with a wear plate which actually runs on the ground. This will, eventually, wear out and should be replaced. The saucer is held on with a single socket screw which can be removed with the Allen key provided with the machine. This socket will, on many soils, get clogged up and will need a knife, nail or screwdriver to clean it out before the Allen key can be used. The saucer should just pull off but may need coaxing with a soft faced hammer.

Some drum mowers have drain holes in the drums. In the C450 they are in the flat-side of each drum, at the bottom. They should, of course, be kept free.

The main drive belts need seasonal attention and may need replacing. Tensions should be more frequently checked. Adjustment is carried out simply by altering one nut, and its lock-nut, on the tensioner so that the right-angled steel tab under the nuts just touches the main sub-frame.

Slacken the break-away off at the beginning of each season and check that it does actually work. Incidentally, damage may result if the mower is lifted off the ground while the break-away is tripped open.

One of the big advantages of rotary mowers is the reduced maintenance requirement. Normal maintenance is concerned with greasing and oiling daily with the occasional replacement of blades. The other items raised here are likely to be seasonal or relatively infrequently required during use. In any case, management of the field operation is likely to be much easier if the machine is kept in good order.

DISC MOWERS

The operation and maintenance of these machines is similar to that required by drum mowers. See pictures in Fig. 6.59.

Fig. 6.59 (a) Ensure that the knives are attached in such a manner that the discs are balanced and that they do not strike the knives of an adjacent disc.

(b) The oil level in the bed of some machines should be checked when the bed is in the vertical position. The life of the bed can be extended if the wearing section is hard faced.

(c) Do not forget to check the gear box oil level. Small oil leaks here often remain unnoticed. This can soon lead to dangerously low oil levels.

(d) The fabric shield must be kept in place when the machine is used.

(e) Keep the belts in line and correctly tensioned.

(f) Various tilt adjustments are possible, but many of them involve changing the length of the top link connections.

COMBINE CARE

The operator's manual is the best guide but to supplement this here are a few faults to look out for and remedy (Fig. 6.60).

Fig. 6.60 Checking the combine.
(a) This vee-belt looked all right, but closer inspection revealed that the cords were showing underneath. Renew a belt if it gets into this state, or if oil or sunlight have perished it.

(b) This is how the drum bars should look. Wire brush them to get all the dirt out and empty the stone trap.

Look for loose nuts and bolts or damaged threads. Examine split pins for signs of wear. They may look small and unimportant but they do an important job.

Go round the machine with a grease gun making sure all the old grease is pumped out. A leaky nipple can sometimes be cured by putting a rag between the gun and nipple to act

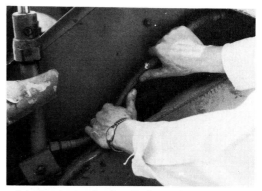

(c) Check over hydraulic hoses for cracks and leaks at the connections and make sure ram seals are sound. Top up the hydraulic oil reservoir.

(d) Set up the cutter bar as for a mower, replacing broken or loose sections. If knife register is incorrect, it is usually adjusted by removing or adding shims at the cutter head drive assembly. Check the retractable tines on the auger. If they are loose, or if any bearings are worn, replace them.

(e) Use a wire brush to clean the sieves and oil the linkages that allow the louvre angle to be altered.

as a seal. If it still leaks, renew it. Make sure sealed bearings have no ruptured seals.

Check sprockets for signs of 'Hooking' —

excessive wear on one side of the teeth — which will wear out a chain quickly. When refitting chains replace the spring clips on the connecting links with the open end facing away from the direction of rotation. Do not overtighten but leave enough play to be able to move the link across the sprocket after tightening.

Free the friction clutches and check the linings. It is also worth rotating the 'clatter' clutches to break any rust seal. These are vital safety checks which could save you a lot of money.

Examine the grain and returns elevators for damaged flights. If a link has to be taken out of the chain, fasten the ends with string to prevent it falling down the box.

Get in a stock of spares such as belts, knife sections, reel tines and lifters and a selection of chain links. If your machine is new this year, the dealer may well have a sale-or-return kit of spares which experience has shown to be the most widely used.

If the combine has warning lights or bells which have not flashed or rung yet, check that they are still working. Make sure the relief driver knows his way round the controls and adjustments of a new machine.

Drum setting

The drum is the heart of the combine. Bad setting and maintenance will give a poorly threshed sample with a low selling price. The main things to look for in a pre-season check are damaged or badly worn rasp bars, drum balance and speed and concave clearance.

First clean out all the dust and dirt. Wire brush the rasp bars to remove caked dirt between the serrations.

Next remove all the drives to the drum and spin it slowly, listening for squeaks or grating sounds that will indicate a damaged bearing. Check that there are no tight spots in the bearings and no sideways and up and down movement.

To check wear on the rasp bars fit a new attachment bolt in an old bar. This will stand proud of the bar and if the bolt protrudes more than 6 mm replace the whole set of bars, because wear will have occurred on all of them. When fitting new bars make sure the serrations face alternate ways.

Checking drum balance is important, because the drum revolves at 1000—1400 r.p.m. and if it is unbalanced the bearings will be ruined and other expensive damage to the drum and concave may occur.

Chalk numbers on the bars of the drum and spin it lightly, noting which bar stops at the top. Repeat three times. If any one bar stops at the top all three times, weight needs adding to that bar to keep it equal in weight to the others. Add weight by sticking flat washers to the bar with grease (Fig. 6.62).

Spin the drum again to check, moving the washers about until the drum is properly balanced. Bolt the washers to the back of the bar and recheck (Fig. 6.63).

If only one bar is badly damaged, through stones or metal getting into the drum, also

Fig. 6.61 Before balancing, identify each blade with a chalked number.

Fig. 6.62 Trial balancing is effected by sticking washers to the lightest bar with grease.

Fig. 6.63 Spin drum to test balance. If one bar stops at the top repeatedly, it should be weighted.

replace the opposite bar to maintain drum balance. Replacing one bar on an old combine will lead to bad threshing because the concaves will have to be set to the new bar and this will give underthreshing of the corn. Narrowing the concave gap will cause the new bar to hit the concave and cause more damage.

Winter storage

Preparing a combine for winter storage is as important as daily maintenance during harvest. A quick rub down, draining the radiator and throwing a sack over the engine are not enough if trouble-free operation is expected next year.

Cleaning, inside and outside, (see pictures in Fig. 6.64) is best done before the muck and dust hardens. Compressed air is the best method, but stubborn dirt may require water or paraffin. Grain and chaff in the elevators and grain pans attract vermin, which look upon rubber seals, canvas flaps and elevator flights as part of their diet. However clean the combine, rats and mice will make it their winter home. Leave inspection doors and covers open so that they may run freely through the machine; if they become trapped they will eat through the first piece of rubber or canvas they come to.

Fig. 6.64 Winter storage procedure.
(a) Example of rubbish accumulation in a neglected machine.

(b) (Left) Remove sieves, wire brush the louvres and lubricate the pivot points.

(c) (Right) Clean the knife thoroughly and coat it with anti-rust. Apply it while the knife is still supported in the cutterbar.

(e) Clean out both cross augers and elevators. Leave the inspection doors off.

(d) Clean out the stone traps and concave.

Running the combine at full blast for a few minutes will get rid of most dirt and dust. Grain in the tank and grain pans may be removed by stuffing a handful of straw into the augers. As the straw is carried along it takes the loose grain with it.

To prevent rust inside soak straw with waste oil, feed it through the drum at a fairly fast rate and run the engine at full throttle. Do this two or three times if necessary.

(f) Release tension on platform lift ram assisting springs and coat the uncovered section of the piston rods with anti-rust.

(g) Wire brush the transmission brake drum and coat with anti-rust. Leave the brake lever in the off position.

(h) Remove straw which has accumulated in the fan housing.

(k) Grease the splines of the pulley drive shaft after levering the flanges apart.

(i) Scrape the inside of the straw walker with a length of pipe.

(l) Jack the wheels clear of the ground and support on blocks. Cover the tyres and reduce the pressure to about 0.7 bar.

(j) Variable speed pulleys should be given special attention to prevent corrosion of the highly polished flanges.

Check nuts and bolts for tightness and grease each bearing, remembering to pump out the old grease. Make a note of unsolved troubles experienced during the season and list worn parts, so

that they may be ordered ready for the yearly overhaul.

Next comes the cutter bar and platform. Remove crop residue and apply a coating of rust preventative to polished surfaces, taking particular care with the cutter bar fingers. Take off the lifter tines and store them in the workshop.

Remove the knife and check it for broken sections. Check the retractable fingers in the feed auger and coat them with anti-rust. Clean off the platform skids and paint the surface with waste oil. Make sure the whole platform is on to a level surface, otherwise it will become distorted.

Slacken the reel drive safety clutch and other friction drive clutches to prevent the discs sticking to the drive plates.

Flat belts should, as a rule, be left in tension to stop them shrinking and twisting, but V belts should be removed or slackened.

For machines with hydraulically-operated speed variator pulleys on the drum and transmission drives the belt tension may be released by starting the engine, moving appropriate to the fastest speed, then stopping the engines and releasing the pressure on the pulley flanges by levering them apart.

The polished parts of all pulleys should be given a coat of anti-rust: keep it off belt surfaces.

Remove chains where possible, wash thoroughly in paraffin and leave in an oil bath. Chains which cannot be removed may be sprayed with a protective compound available in aerosols.

Retract all hydraulic rams or coat the uncovered sections of the piston rod with thick heavy grease. The rams and shaft splines in the speed variator pulleys should also be treated. Care must be taken not to get grease on the belts.

External brake drums and discs should be cleaned and coated with anti-rust.

Jack up the wheels and reduce the tyre pressure to about 0.7 bar. Release the brakes and fix the clutch pedal in the disengaged position to prevent sticking and seizing during the winter.

Check the paintwork and touch up damaged areas.

Winter storage of the engine (see Fig. 6.65)

Run the engine until it reaches normal operating temperature, then change the oil and oil filter.

Clean the air filter and change the oil. Run the engine for about 15 minutes and, while still

Fig. 6.65 Winter storage of the engine.

(a) Dirt on the battery terminals tends to collect moisture. This can cause a direct short between two cells.

(b) An example of frost damage to a cylinder head.

(c) An old washing-up detergent container makes an excellent oil can.

(d) Seal the exhaust pipe and all air intake pipes.

running, squirt upper cylinder lubricant into the air intake manifold. Stop the engine while it is exhausting blue smoke.

Seal air intake pipes, crankcase breather, exhaust pipe and oil filter with waterproof material (polythene bags are ideal). Cover the dynamo and control box.

Drain the radiator and put the filler cap on the driver's seat as a warning.

Top up the fuel tank to prevent the build-up of condensation and seal the filler cap.

Remove the battery, keep it topped up and charged; if possible, use it on another vehicle for a few days every two months.

FORAGE HARVESTERS

To prepare the harvester for work go over each grease point making sure that the new grease forces out the old, then wipe the points clean. Check the gear box oil and top up. Remember to look again every 50 hours and drain and refill every 200 hours. Wash chains in paraffin, replace and lubricate with SAE 90 gear oil applied on the top side of the lower chain span. Adjust to allow 10 mm slack for each 300 mm between sprockets. If it is too tight it will overload the bearings; if too slack it may jump off a sprocket or mount a tooth and break. Belts should be well tensioned to transmit the considerable power. Allow no more than 20 to 30 mm up and down movement between pulleys (Fig. 6.66). Slip clutches tend to stick after storage. Free them by slackening the adjustment bolts and turning unit by hand until the clutch slips. Then retighten each bolt the same amount until the clutch does not slip under load. Inflate tyres to recommended pressure.

Fig. 6.66 Belt tension check. Note 20 to 30 mm play between pulleys but no more.

Broken or damaged flails must be replaced, as should any worn down more than 10 mm or they will not create sufficient wind to blow material up the chute. An equal number must be put on each rod to maintain balance. Flail tips can usually be sharpened by grinding. Hold the flail against the grind wheel so that it is ground back about 8 mm at an angle to give a sharp edge (Fig. 6.67). Take the same amount off each flail to keep the rotor balanced. The shear bar is usually adjustable. Set the clearance between it and the flails according to the degree of laceration and length of cut required — 18 mm from the tips of the flails for beet tops and 6 mm for grass. Check this clearance periodically

Fig. 6.67 Close-up of a sharpened flail.

and as the flails wear down, adjust the shear bar accordingly.

For safety do not stand in front of a flail harvester particularly when a low cut is being made where there are a lot of stones about.

On the double-chop type forager with cutting flails and flywheel mounted knife cutters, the same flail procedure should be followed to check condition and sharpness and replace any broken, bent, or badly worn. Check retaining bolts daily and tighten to 240–250 newton metres of torque. The knives and paddles on the cutting head must be equally spaced — two knives with two paddles and so on. Each knife is usually adjustable to the shear plate by loosening and tightening two nuts. Adjust to 1.5 mm from the plate and tap each bolt head with a hammer to ensure it is seating tightly and tighten both nuts securely.

Set paddle clearance between tips and casing 3 to 4.5 mm. If a knife is bent it can be removed and hammered out but do not strike the specially hardened edges. Shear plate must be kept sharp and adjusted as blunt cutting edges and excessive clearance waste power and mangle the crop.

Never inspect or adjust while the machine is in motion. Turn mechanism by hand to see that it is free-moving before running it under power.

The cylinder chopper is similar in action to the knife cutting unit of a lawn mower. The knives must be spaced equally throughout the cylinder otherwise the cutterhead will be out of balance and uniform length of cut will be impossible. The number of knives decides the length of cut and when changing them the knife support as well as the knife must be removed. Knives must be kept sharp and shear bar adjusted to 0.04 mm from them. Failure to maintain these will give poor quality cutting, raise power consumption and increase wear of both knives and shear bar.

When a stone knife sharpener is incorporated it must be held against each blade and moved along its length while the pto drives the cutting cylinder at about ¼ speed. Then adjust the shear bar to the sharpened knives after removing it and clearing out material which has collected there. Wear safety goggles during sharpening.

Forage harvesters are power consuming. The faster a flail rotor turns the greater is the demand upon the tractor engine. The two must be matched so that the recommended rotor speed is achieved at the tractor's rated speed. The size of tractor required depends upon the width of cut, forward speed and crop density. Rotor speeds may vary between 1000 and 1500 r.p.m. and the tractor and harvester should have some power in hand in order to maintain output under varying operating conditions. The gear which gives the required pto speed without labouring the engine should be chosen and any inequalities in crop density overcome by taking a narrower cut rather than by reducing forward speed. The height of cut of most machines is adjustable from about 25 to 150 mm and the harvester must be set so that it is level in both directions. Skids are there to protect the rotor should the wheels drop into a hollow and these should be set to give a clearance of between 50 to 75 mm between them and the ground.

The forage harvester's rotor must be properly balanced, otherwise it will consume extra power and damage the bearings.

On many models the rotor is dynamically balanced before the machine leaves the factory, to give smooth, easy running and prevent excess wear on the rotor bearings.

Fig. 6.68 Two flails — one properly reground, the other well worn.

Broken or missing flails are common causes of poor balancing, so before the season begins check the rotor and flails (Fig. 6.68).

Most rotors have the flails in banks or rows. Remove one row at a time and inspect for loose or missing hardware, bent or broken flails and cracks.

A replacement flail must be hacksawed back to the length of the others or it will cut a different length. Some flails wear more than others and some at a different angle. Grind them all to within about 3 mm of each other. The edge, when sharp, should be parallel to the ground when the flail is hanging vertically from the rotor. Too long an edge will give a weak blade tip that will easily break off (Fig. 6.69).

Fig. 6.69 (a) Grinding a flail to the correct cutting angle.

(b) An example of inconsistent grinding (left) and an overlong edge (right) compared with a perfect flail in the centre.

Before attaching the flails to the rotor, inspect the bushes on which they rotate for splits and wear, and also the bolt that holds the complete flail to the rotor. The flail should rotate around the bush, but often the bush rotates round the bolt, wearing out the bolt. This is caused by not tightening the bolt sufficiently.

Finally take out the shear plate and sharpen it on the grindstone. Set it in the field, according to length of chop and laceration required. Check belt tension and alignment and gearbox oil levels. And lubricate all grease points.

BALER OVERHAUL

If the pick-up baler was not thoroughly checked before haymaking, look it over before harvest. If it is to stand idle for a time, protect shiny parts and pull out material left in the bale-chamber. Before using the baler trip the knotters and turn the baler over slowly by hand to make sure all moving parts are free. Check that the pto shafts slide easily and the guards are in position. Look at the universal couplings for cracks or missing dust caps.

The flywheel slip-clutch is the primary safety device of the baler. It protects the machine from gradual overload and prevents serious damage.

Its principle is to limit the amount of torque supplied by the pto. If it slips too soon the baler output is reduced, and if it does not slip overloading of the baler results and a twisted bale chamber may arise, causing a long delay.

The lay-out of a typical clutch is that the machined surface of the driven member is trapped by two circular friction linings between spring-loaded pressure plates (Fig. 6.70). The torque input to the baler is governed by the compression of the individual springs. Most manufacturers recommend in their instruction booklets a minimum length the springs should

Fig. 6.71 Worn slip clutch components. Foreground, drive plate dogs sheared off. Background, badly-worn dogs should be replaced.

Worn clutch plates — one with the centre twisted out.

be compressed and this figure should be closely followed.

Main causes of trouble to check are broken springs, rusted linings, oil or grease on the linings (Fig. 6.71). To inspect these parts properly the unit should be dismantled by removing the over-run clutch first, and cleaned. If the driven member is badly damaged it will need renewing as will the linings and perhaps the springs (Fig. 6.72).

After servicing and replacement of worn or damaged parts take care to reassemble correctly, tighten each nut up on the springs equally and fit the unit back on to the flywheel.

Next set the torque to the manufacturer's recommendation figure, e.g., several Sperry New Holland models require a torque of 270 Nm.

To check clutch action, place a wrench on

Fig. 6.70

Fig. 6.72 Clean up pressure plates with emery cloth to remove rust and pitting, and check for damage.

the baler pto (power take off) shaft and lock the flywheel or the plunger. If the clutch is operating properly a force of 45 kgf (100 lb) applied on the handle of the wrench at a point 600 mm (24 in) from the centre of the pto shaft should cause the clutch to slip. If the clutch slips too easily tighten each nut equally in turn until the correct torque is obtained (Fig. 6.73).

A final check should be made after the baler has been in work for a while to ensure that the clutch is not unduly hot.

Fig. 6.73 Check slip-clutch torque with a spring balance on a bar 600 mm from the pto joint.

Sharpening and resetting knives

The knives in the bale chamber work hard, making about 70—80 cuts per minute. Before the baling season they should be removed and sharpened on the grindstone.

Take care when sharpening not to over-heat the blade or it will lose its hardness and quickly become blunt when put to work. Grind off a little at a time, cooling the blade frequently in a bucket of water.

Keep the blade angle as near as possible to that of the original and keep the edge straight. Small notches in the blade caused by stones or wire do not matter (Fig. 6.74).

Fig. 6.74 Damage caused to a knife by stones entering baler.

Large chips from the knife edge can be filled in only by welding and resharpening by a specialist.

Some balers have a shear bar instead of a stationary knife. This should be removed and sharpened as its shearing edge will become rounded after a season's work. Grind it back to its original profile.

Next refit the stationary knife or shear bar and set the ram-knife parallel to it until there is about 0.8 mm ($^1/_{32}$ of an inch) clearance. This clearance is critical. Too much will give a rough-edged bale and consume more power and too little will cause the knives to hit each other.

While removing and refitting the knife and checking the clearance, wedge the flywheel to prevent it turning or get someone to hold it, otherwise the ram may move along the bale chamber and cause serious hand injuries.

Shims are used to help set the clearance.

Fig. 6.75 How to position shims in relation to the knife.

They are thin strips of metal placed behind the knife. By adding a shim or taking one away, the clearance can be set accurately (Fig. 6.75).

Two types are made for baler knives. One is a long shim that fits the length of knife and the other is a small, short shim used at the ends to ensure that it is parallel with the other knife.

To get an accurate setting it may be necessary to add and subtract shims. Check several times, tightening up the screws each time. Check clearance at each end of the knife with feeler gauge (Fig. 6.76).

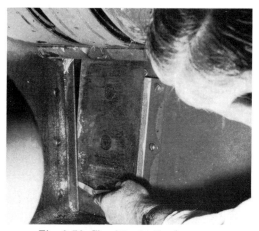

Fig. 6.76 Checking knife clearance.

Look for wear on ram bearings and runners and adjust the ram centrally in the balechamber. This alignment of the ram is one of the most important basic adjustments on the baler. It is also often completely neglected. Correct bearing

adjustment is needed to ensure that the ram runs parallel to the bale chamber with the correct clearance or 'play' on the runners at top and bottom over the full length of travel. Excessive or uneven play causes increased power consumption, increased vibration and may upset the smooth running of the rest of the baler. See that the ram-stop enters the bale-chamber when the needles are in.

Inspect the knotters for wear on the bill-hooks, retainers, twine-fingers and cam-bearings, and see that the twine knives are sharp. The number of stoppages caused by knotter break-downs is far greater than in any other part of the baler. Look at the needles for a groove where the string is pulled through each time it enters the balechamber. Braze the needles as shown in Fig. 6.77 and they will last another season. Use a wirebrush on the knotter brake disc to remove all rust and oil.

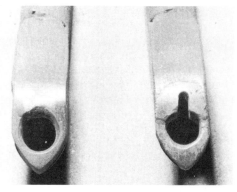

Fig. 6.77 Worn and repaired needle.

Oil the bale tension adjuster threads and check the springs for cracked or broken coils. Have a look at the feeder mechanism for faulty bearings and tines. Turn the pickup and replace any broken or badly bent tines. Keep a few spare tines handy as these often snap during the season.

Check all oil levels and tension chains correctly. Go round the machine with a grease-gun removing any blocked nipples and cleaning or replacing them.

Check tyre pressures and see the bale counter has not seized. Renew oil in the gearboxes, and lay in a stock of correct shear-bolts and string.

Run the baler up to speed and bale some dry straw to remove rust and any stiffness from the ram and balechamber.

Shear bolt sense

By replacing the manufacturer's specified fly-wheel shear bolt with a stronger one you may increase the performance of your baler. But damage to the balechamber, gearbox, needles or ram could occur because of overloading.

Shear bolts are made to fail at a load determined by the designer. They protect the rest of the machine, so always use the manufacturer's shear bolts.

Buy sufficient to last a season and see there are some in the tool-box.

The shear bolt insert in the flywheel is hardened, but will eventually wear and become a loose fit. The bolt may then shear before it should, giving a decreased output from the baler. Before each season check the wear on the insert with a new shear bolt. As a guide, if the bolt hole is noticeably oval the bolt should be replaced.

Before refitting a new one check it and the hole it fits for burrs and remove all dirt from both parts. Use a parallel punch with a large shoulder to fit the new insert. A block of hardwood will suffice if no punches are available. Do not use a taper punch, which may burr the edges and cause the insert to become jammed (Fig. 6.78).

Tap in the insert until it is flush with the front face of the flywheel. It should not protrude or it will take some of the load instead of the shear bolt. Always use two ring spanners for tightening shear bolts, to avoid injury.

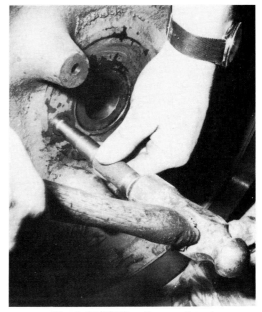

Fig. 6.78 Fitting the new insert.

One last time-saving tip: carry a small punch and hammer to knock out broken shear bolts.

Tension all chains correctly, check the pick-up reel for broken or missing tines, inspect ram bearings and runners for wear, and grease the machine regularly.

In-the-field cures

To many of us baler knotters are a mystery and until they go wrong they are best left alone,

Fig. 6.79 (Photograph copyright by Peter Adams).

otherwise they may be incorrectly adjusted and give constant trouble.

NIAE user test reports show that correctly set up knotters will give little trouble. In the chart on the next pages are some of the most common causes of knotter breakdown and some of their remedies, all of which can be carried out by a skilled tractor driver.

It will pay to have knotters checked over by your dealer just before the baling season begins.

Remember a baler will still function, though inefficiently, with a blunt shearing knife or incorrectly set slip clutches, but if the knotters are wrong it will not begin to bale.

Fault	Cause	Cure
1. Twine cut or broken without a knot being formed.	(a) Insufficient twine being drawn through retainer to form knot. (b) Twine cut by sharp edges on retainer.	(a) Clean retainers of dirt. Adjust tension on retainer spring. (b) Smooth rough edges on retainer.
2. Long end on one side of knot.	(a) Blunt twine knife. (b) Knife arm has insufficient lift to cut both twines.	(a) Sharpen. (b) Carefully bend knife arm casting to increase lift on knife arm.
3. Twine tangles or breaks in spools.	Incorrect threading.	Re-thread baler.
4. Twine breaks on release from bale chamber.	Excess bale-tension.	Slacken bale-chamber tension.
5. Twine slips off one side of the bale.	(a) Packing of bale not even. (b) Bale-chamber tensioners unevenly adjusted.	(a) Adjust packer fingers to give even density bale (Fig. 6.80). (b) Adjust tensioners to give equal tightness on both strings.
6. Both ends untied.	(a) Twine not held tightly enough by knotter-hook jaw for knot to be formed. (b) Knotter-hook tension inadequate.	(a) Strained hook-jaws. Replace. (b) Increase tension (Fig. 6.81).
7. Knot stays on knotter hook.	(a) Too much tension on knotter-hook. (b) Knife too close to retainer face. (c) Stripper arm not lifting knot off hook.	(a) Ease off knotter-hook tension. (b) Increase distance between knife and retainer face. (c) Bend stripper to lightly touch knotter-hook (Fig. 6.82).
8. Knot only on retainer end of twine.	(a) Needle not laying twine over knotter-hook. (b) Twine too slack on leaving spools.	(a) Adjust needle travel (Fig. 6.83). (b) Increase twine tension out of twine box.

Fault	Cause	Cure
9. Loop knot.	(a) Twine-knife blunt.	(a) Sharpen twine knife.
	(b) Twine retainer tension too weak.	(b) Adjust twine retainer tension (Fig. 6.84).
10. Shear bolts breaking continually.	(a) Fly-wheel clutch incorrectly adjusted.	(a) Check clutch adjustment (Fig. 6.85).
	(b) Wrong grade of shear-bolt.	(b) Fit recommended bolts.
	(c) Bale-chamber tension excessive.	(c) Ease off bale-chamber tension springs.
	(d) Incorrect pto speed.	(d) Run tractor at recommended pto speed.
11. Rough edged bales.	Knife incorrectly adjusted or blunted.	Sharpen knife and adjust clearance.
12. Crop left on ground.	(a) Pick-up height incorrect.	(a) Adjust.
	(b) Tines missing or broken.	(b) Replace.

Fig. 6.80 Adjust packers to give correct bale shape.

Fig. 6.82 Bend stripper arm to lightly touch the knotter hook.

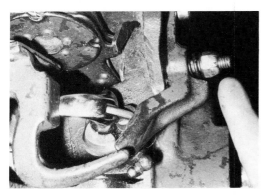

Fig. 6.81 Knotter hook tension adjuster.

Fig. 6.83 Needle travel adjustment.

Fig. 6.84 Twine retainer tension adjuster.

Fig. 6.86 (a) Always support the rear gate on its safety latch whenever an internal inspection of the machine is made.

Fig. 6.85 Slip-clutch adjustment springs.

(b) Set the pick-up height as high as possible without leaving any crop on the ground.

THE BIG BALER

There are many types of big baler on the market. Most of them produce round bales and are capable of baling a variety of crops.

It is important to obtain even windrows. With some machines it is advisable to drive a zigzag course over narrow windrows in order to obtain evenly packed bales.

Round bales are generally more weather-proof than convential bales and can thus usually be left outside in the winter. Another advantage is that they will continue to dry after they are baled.

Although the machine should normally be run at standard pto speed, in certain circumstances improved feeding and rolling can be achieved if the reduced pto speed is used. The lower speed will result in less crop damage and thus the formation of less trash, which may otherwise damage belts or rollers.

Some of the main points relating to the operation of a big baler are shown in Fig. 6.86.

(c) Note that many machines require two sets of double spools in order to operate their hydraulic components.

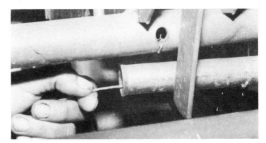

(d) Threading the machine may involve pushing the twine through a long narrow tube. A length of welding rod will help with this job.

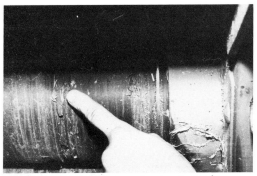

(e) Belt or roller damage due to stones is fairly common. Sometimes the friction caused by belts rubbing against each other or a stationary build-up of crop can lead to the machine catching fire.

(f) Changing the baler density is normally only a matter of changing the tension on a spring-loaded arm.

Winter storage

Before winter storing the baler, first clean it inside and out to remove dirt, old grease and crop residue (Fig. 6.87(a)). Dirt left on surfaces over winter gathers moisture, which

Fig. 6.87 Winter storage procedure.
(a) This tractor exhaust cleaning attachment can be used for applying a protective coating of oil as well as removing dirt.

(b) Clutch friction surfaces are best left without any special treatment. Apply oil to the nuts and stud threads.

causes corrosion and in time may cause paint to peel. It is pointless to treat with anti-rust surfaces which are dirty.

Disconnect the pto shaft safety slip-clutch from the flywheel and slacken off the spring tension (Fig. 6.87(b)).

Remove the flywheel shear bolt and coat the drive flanges with thick grease (Fig. 6.87(c)). Leave the shear bolt out, to prevent children turning the baler and injuring themselves.

(c) Grease will prevent the drive flanges seizing together.

Check the pto shaft safety guards and grease the universal couplings, taking care not to pump in too much grease and rupture the dust seals.

Wash down the knotters and remove any strands of twine from the retainers. Clean dirt from the teeth of drive gears (Fig. 6.87(d)) and, once the knotter assemblies are clean, coat them with oil or anti-rust, grease all nipples and protect with a waterproof cover.

(d) The knotter disc brake should not be oiled or greased. Either leave it dry or use a special anti-rust solution.

Chains should be cleaned with paraffin and coated with thick grease (Fig. 6.87(e)). Some must be removed for treatment and care must be taken when refitting to see that the timing is correct. Another way of treating chains is to boil them for half-an-hour in transmission oil or tallow. This allows the lubricant to penetrate into the pins and bushes.

Clear any crop which has become wound

round the pick-up tine bars and clean the drive cam (Fig. 6.87(f)). Coat with anti-rust the stripper loops and all other areas where paint has worn off.

(f) Removing the end stripper loops allows easy access to the pick-up drive cam.

(e) Soak chain in paraffin for half an hour then brush off old grease and dirt.

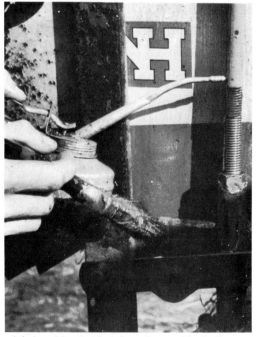

(g) An old paint brush makes an ideal oil applicator and ensures that the whole area is covered.

(h) After applying grease to the plunger guide rails, turn the baler over a couple of times to distribute it.

To treat the bale chamber, first unscrew the bale density adjusting cranks and oil the adjuster screw threads (Fig. 6.87(g)). Pull out the last couple of bales. If these are left over the winter they will gather moisture, expand and could buckle the sides of the chamber. Coat the inside of the chamber with oil and cover the plunger guide rails with thick grease (Fig. 6.87(h)). Soaking straw with waste oil and feeding it through the baler makes this job easier.

Remove both crop knives, resharpen, coat with grease and store in a dry place (Fig. 6.87(i)). Coat twine tensioners with grease (Fig. 6.87(j)). Rub down and repaint all chipped or scratched surfaces, using a red oxide primer and an enamel topcoat.

Jack up the baler on blocks.

(j) Coat twine tension plates and guides with grease.

BEET HARVESTER OVERHAUL

Web chains, rollers and couplings

Preparing a beet harvester for its season's lifting is not as involved as preparing more complicated machines such as combines and balers. Nevertheless, a few hours spent on checking, adjusting and greasing will pay dividends when the going gets tough. There are few machines that have a tougher life.

First overhaul the web chains. Check each

individual link for wear at the point where it comes in contact with its counterpart and replace any that are severely worn. Links worn more than half their diameter will probably fracture during work and it is far easier to change them in the workshop than in the field when the chain is covered in muck. Figure 6.88 shows a web chain being joined.

Fig. 6.88 Joining a web chain.

(a) Joining a web chain the easy way: first attach one link to the two loose ends of the chain and, using it as a lever, pull the two sections together.

(b) Next, hook the link on to the lower section. Pull links as close as possible by gripping with both hands and connect.

(c) Final hooking up is thus confined to one end. Even with a new, tight chain, one hammer blow should complete the joint.

Repairing links by welding is not really practicable. Welded links tend to break off either side of the weld, but if in an emergency a weld has to be made ensure that both ends of the link are the same diameter after welding. If they are not, the chain will run out of line or snag on the drive sprocket.

Sprockets often become hooked causing the chain to hang on or jump off. Some sprockets

Fig. 6.89 An example of a badly worn and hooked sprocket. The profile of the teeth has been changed by the amount of wear shown at A. This will cause excessive strain in bearings, shafts and chains.

Fig. 6.90 These two roller chains are completely ruined through lack of attention. The majority of the link pins are seized solid and rollers are caked in soil and rust. They should have been removed and left to soak in an oil bath.

Fig. 6.91 Worn guide rollers, if not attended to, will increase the rate of wear in the web chain links. Remember these types of bushes must not be greased or oiled.

may be reversed, but this should only be attempted if they have not been allowed to become more than half worn (Fig. 6.89).

Adjustment of sag for a web chain will depend on whether it is running horizontally or at an angle; normally it should be just tight enough to allow easy joining. If the going is sticky, a slightly slacker chain will give a better cleaning action. Don't forget the roller chains (Fig. 6.90).

Chain idler guide rollers should be free running. Check that they have not worn on one side only. Most manufacturers mount these guide rollers on chilled cast bearings which have glass hard bearing surfaces. These should not be greased or oiled as they are not

sealed to keep out dust and dirt. Grease and soil when mixed together, make an excellent grinding paste (Fig. 6.91).

Sliding couplings on pto and other drive shafts tend to seize or become sticky during winter storage. To prevent damaging the universal joints through excessive end thrust clean off all the old stiff grease by washing in paraffin and smear with fresh grease. When pumping grease into the joints, care must be taken to avoid blowing out the dust seals; one or two pumps are usually sufficient.

A squirt of release oil on all bolts and set screws used for adjusting will save time and knuckle damage in the field.

The slip clutch should be dismantled,

inspected for wear, cleaned and reassembled dry. On no account put oil or grease on the friction surfaces and do not overtighten.

Topping units and flails

The setting of the beet harvester's topping unit will to a great extent determine how much beet is lost or wasted. As a guide keep the knife flat in all directions, have the feeler wheel centred on the crown of the beet and remember that the amount of crown to be removed is governed by the distance between knife and feeler wheel, if tops are long and

lifting rate of a harvester. Check that serrations on the feeler wheel are sharp, the assembly free to float, the knife sharp, and the knife breakaway device working.

Rubber flails take a pounding during a season's lifting, especially in stony conditions. Check flails for fractures where they bolt to the hub and replace any that are suspect. It is often wise to replace the complete set of flails to ensure the spinner is balanced. At the same time, take a look at the drive chain and sprocket — shock loads transmitted by flails hitting stones tend to stretch the links and wear the sprocket teeth.

Flails can do more harm than good if not

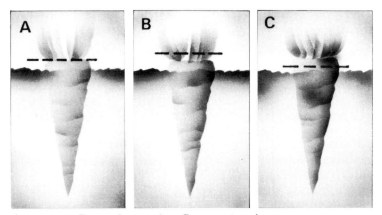

A — correct; B — under topping; C — over topping.

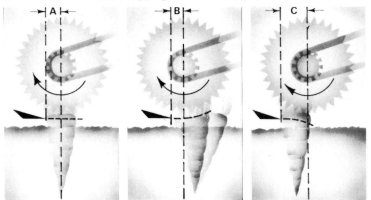

Fig. 6.92 A — correct; feeler wheel grips the beet all the time during cutting. B — cutting too early; knife cutting before the feeler wheel grips the beet. C — cutting too late; feeler wheel running off the crown before the cut is completed.

sturdy this distance needs to be greater than for small, drooping tops (Fig. 6.92).

A return of 'no top tare' means that too much crown has been removed.

Efficiency of the topping unit governs

set in the correct position. For normal setting they work best just touching the ground, but on light, sandy soil they need to be slightly higher, to prevent beet being knocked out of line and missed on the next round (Fig. 6.93).

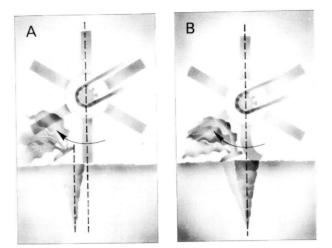

Fig. 6.93 (A) Correct setting; flails slightly offset and just brushing the ground. (B) Incorrect setting; flails set directly over the beet will not throw the tops the required distance. If too close to the ground beet will be dislodged and also cause damage to the flails.

Adjust the trash disc to cut off enough of the beet top to allow the knife arm a clear run. Failure to do this will result in trash build-up on the arm and blockages.

To overcome crabbing on sliding ground, move the trash disc into the centre of the two so that it cuts a furrow deep enough to guide the harvester wheel. Although this will cause a build-up of trash around the topping unit and knife arm, it is probably the lesser of the two evils.

Share width setting

Having got the topping part right, turn to the lifting mechanism.

For machines with lifter wheels instead of shares it may be necessary to reset the tapered roller bearings. First remove the grease cap and the cotter pin from the castellated retaining nut, tighten the nut until the wheel is stiff to turn, back off the nut one flat or until a slot lines up with the hole in the axle and re-fit the cotter pin. Fill the grease cap with the recommended grease before reassembling.

Lifter wheels, like shares, don't last for ever. When it becomes necessary to change them don't throw away the old set; they may come in handy in an emergency. Worn wheels do not block up as quickly as new ones and some contractors carry a set to get out of trouble in heavy and sticky conditions.

Lifting shares, if caught before they have

worn too far, may be built up with weld. The lift of a share can be greatly increased by welding a piece of spring steel on to the heel, taking care to retain the original profile.

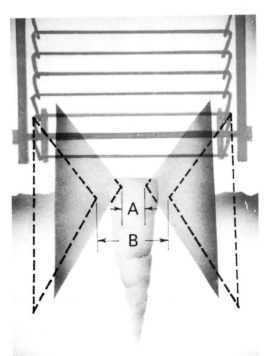

Fig. 6.94 Share width setting is governed by the size of the beet; the average setting is about 40 mm. A is correct for the beet shown in the diagram, but B is wide and would leave a lot of beet in the ground.

When working hard ground new shares often break off at the point. Fitting a worn set may overcome the problem. The average share width setting is 40 mm, usually obtained by adding or removing shims. Mounting bolts must be really tight to prevent the shares moving when in work (Fig. 6.94).

How much soil should be lifted with the beet? This is a case of setting to suit conditions, but at all times the beet should be lifted with tap-roots intact.

In light soil it is advisable to take a little more soil than usual to safeguard against losing the smaller beet, but remember that the factory will reject any load that has an excessive amount of soil.

Field operation

Some harvesters are fitted with an attachment for opening out a field (Fig. 6.95) but too often this back-breaking job has to be done by hand. Sharp shovels and dutch hoes are the best tools for this job.

After the field or fields have been prepared for the machine, it is good practice to do all the machine opening at one go. This saves having to readjust wheel widths each time on entering a new field. Open out during the most favourable ground and weather conditions; it saves beet and simplifies what is sometimes a difficult operation.

The correct wheel setting for tractor and harvester is important and it is best to have it right from the outset. Tractor wheel centres should be three times the row width; for example, the setting for 500 mm rows would be 1500 mm centres.

The harvester wheels during opening out have to be centred between the rows, but for normal operation the right-hand wheel — looking from the rear of machine — should be centred between the rows and following tractor wheel, whereas the left-hand wheel should run in the bottom of the furrow made by lifting the previous row. This helps to keep the harvester stable and follow the drills, especially on sidling ground.

Fig. 6.96 Adjusting feeder flails to strike the edges of lifter wheels minimises soil build-up.

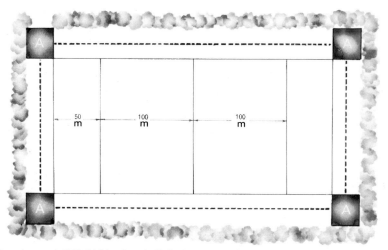

Fig. 6.95 Opening out. Lift 'A' by hand. Take harvester down the middle row of headland and lift beet towards the hedge. This, usually the wettest part of the field, should be lifted first. Then drive back lifting towards the centre of the field to remove rest of the headland. Divide remainder into lands (50 m is normal). Over-large lands will increase out-of-work running on headlands.

Moving the hitching position of the machine on the linkage drawbar instead of adjusting the wheels will result in either the harvester or the tractor crabbing. This can cause damage to the beet through the wheels hitting them or, even more troublesome, disturb the firm soil at the base of the beet and make it difficult for th topping unit to work efficiently. Adjust the feeder flails to strike the edges of the lifter wheels (see Fig. 6.96).

Carry a selection of spare parts. A first-aid kit comprising a few web chain links, roller chain links and half links, a couple of lifting shares, a set of rubber flails and two sharp knives will help in an emergency.

DRILL OVERHAUL

Many drills will be brought out in the middle of March for the first time since last season. A little attention to correct setting up and regular daily maintenance could save money, time and breakdowns.

If the drill was put away properly at the end of last season the hoppers should be clean and free of corrosion. Feed mechanisms, slip clutches and wheel bearings should also have been checked last year — now is not the best time to be chasing spares.

The movable fertiliser feed parts — star wheels and their gear assemblies — should be removed from their winter storage in creosote or diesel oil and refitted to the hopper.

The feed spouts to the coulters should be rechecked — for holes and splits in the case of rubber and plastic and corrosion and distortion on the metal type (Fig. 6.97).

If a large area is to be sown with one particular type, variety or batch of seed, it is worth calibrating the machine to make sure that the correct amount of seed is being drilled.

Many farmers argue that this operation is unnecessary, but a practical example speaks for itself. A Lincolnshire farmer using the manufacturer's recommended setting on a large batch of seed for a 40 hectare piece of land decided to check. His calibration showed that the machine was sowing over 60 kg a hectare too much. Almost 2.5 tonnes of seed were being wasted on the whole operation.

The ideal way to calibrate is with bags attached to the out-of-work spouts while running in the field to be drilled, but fields vary and a proper stationary test gives a good basic guide. Our picture sequence (Fig. 6.98)

Fig. 6.98 Calibrating a drill.

(a) The number of wheel turns necessary to travel a tenth of a hectare is calculated using the formula:

$$\frac{1000}{\text{sowing width (m)} \times \text{diameter of wheel (m)} \times 3.14}$$

Mark a wheel spoke and the drill before turning the required number of turns.

(b) Many modern drills are now equipped with a handle and calibration tray which can be used without having to jack up the drive wheel of the drill.

Fig. 6.97 Seed and fertiliser feed spouts, whether metal or plastic, should be checked for damage before use.

(c) The seed collected is then weighed. This
weight multiplied by 10, plus an addition of
10% to allow for wheel slip, will give the rate
per hectare. As the amount of wheel slip is not
known during this static test, the only way to
really calibrate is to tie a plastic bag over
several seed spouts and drive up and down the
field where drilling is to actually work. The
weight of seed gained, multiplied up, will
give the actual rate per hectare.

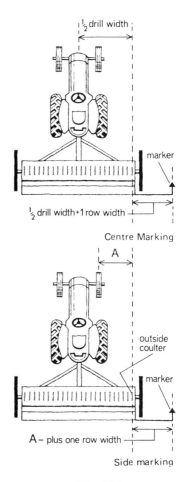

Fig. 6.99

shows the method. Remember, however,
that tests at Edinburgh University show that
wheel slip on the drill in the field may be over
20%.

Most drills have a drive gear for high seed
rate and low seed rate gear to the feed mechanism
and the gear box and finer feed settings step the
rate up and down in 7 kg stages.

To avoid wasting seed and land, markers
must be set up correctly. Some drivers prefer
centre marking, others side marking; our
diagrams show the setting method for both.

For trouble-free operation all disc coulter
bearings should be greased twice daily and all
the other grease points once a day. Aim to
force grease into the bearing until the dirt
laden grease is forced out.

To keep the fertiliser side free of trouble
either fill the hopper at the end of the day's
work or empty it and brush it out. The first
method is safe if you are confident that the
weather will allow you to get back to the job
next day but use the second if the weather
looks threatening, otherwise you will have

Fig. 6.100 Daily maintenance is obviously
important on the drill; of particular importance
is the smooth and even functioning of coulters.

a box full of sludge.

The seed side of the machine should not give much trouble as long as you keep the seed bag labels and pieces of paper bag out of the hopper. Check daily that both seed and fertiliser are getting into the ground.

SPRAYER OVERHAUL

The success or failure of chemical sprays depends on how expertly they are applied. As little as 5 grams of chemical may have to be placed evenly over one hectare and to spray without thoroughly inspecting the machine can result in a valuable crop getting a killer dose or none at all.

Calibration is all-important. First check the handbook for the correct nozzle tips for the application rate and the pressure at which they should be worked. Set the engine speed to give, say, 6.5 kmph in the gear to be used — check that this is achieved in practice by measuring the distance travelled in one minute in the field. If the tractor has a speedometer, so much the better and easier.

If you have a tramlining system and are, for example, working a 12 m sprayer behind 3 runs of a 4 m drill, then the calibration of the sprayer is worked out like this:

There are 10000 sq metres in a hectare and 6.5 kmph is 6,500 m/hr. So, for a 12 m sprayer to cover one hectare, it will take:

$$\frac{10000 \times 60}{6500 \times 12} = 7.69 \text{ minutes}$$

Fill up the sprayer tank, run it for 7.69 minutes (i.e. 7 min 41 sec) and measure how many litres are required to refill the tank but remember that the tractor engine speed must be the same as that required to run the tractor at the chosen ground speed in the field (i.e. in this example, at 6.5 kmph). To find the amount of spray that would be applied to one hectare measure the amount of water required to refill the tank.

Method is as important as the measure. Spray twice round the outside of the field to give a good headland and where possible spray the remainder, following the drills. Better still, replan your operation as soon as possible to run on a tramline system. If this is not feasible yet, take some care in marking out.

Tie a length of cord 2 m longer than the boom to each end so that when turning at the headland the free end of the cord will stay in position until the turn is completed, making it easy to match up the next run.

Watch for blocked nozzles and steer a straight course. If a nozzle becomes blocked do not stop in mid-field, otherwise the drips from the nozzles may damage the crop. Stop the sprayer and, using the previous wheel marks, drive to the headland. Reverse into the hedge bottom and put on a set of clean tips.

Wash and clean the sprayer at the end of each day's work, bearing in mind that it may be a week or so before it is next used.

Don't forget the pump just because it is hidden under the tank. Clean and lightly lubricate the drive coupling and make sure it can be turned freely by hand before coupling on the top shaft. Inspect the hoses, keep the face-bolts tight and make sure the securing chain is fixed.

Wash out the tank to remove any scale formed during storage, especially under the tank top. Inspect the filters for breaks in the gauze and repair minor splits with solder. Filter inserts should be changed according to the manual and a check on the jubilee clips with a screwdriver is useful (Fig. 6.101).

Fig. 6.101 Inspect the filters for breaks in the gauze.

Nozzle care

Remove all the nozzles, filters and end caps and wash thoroughly (Fig. 6.102). Pump water through and tap the pipes at the same time to dislodge any foreign matter.

Clean nozzle filters, but do not over-tighten when re-fitting. It is best to fit new tips to make sure the pattern is right from the outset. When replacing the nozzles keep the pump working and, starting from the centre, work outwards, so that as each nozzle is fitted the pressure in the boom builds up, flushing out

Fig. 6.102 Filters and nozzle.

Fig. 6.103 Demonstration of how some chemicals will not mix with water unless agitated.

dirt through the ends. Check that the end caps are a good fit.

Spray water through the nozzles and inspect each one for correct delivery. Even if the slightest streak can be seen, change the tip — remembering that even new tips can be faulty. A simple check is to hold a jam jar under each nozzle for a set period. Unless the nozzles are right the rest can be a waste of time.

Some chemicals do not mix easily, so half fill the tank with clean water, start the pump and agitator and then add the chemical. Never put neat chemical in the tank or it will go straight down the pipes and not be mixed. Wash everything used during mixing, top up the tank and stir with a clean stick (Fig. 6.103).

To keep foaming to a minimum, check the suction pipes for air leaks and if the sprayer has a recirculatory agitating system make sure the end of the return pipe is always covered with liquid.

Finally, make sure the pressure gauge is working correctly. It gets a lot of vibration on a mounted unit and it is advisable to change it every year.

Precautions

Having set up the sprayer there are a number of do's and don'ts about using it which must be followed. Not the least of these are adequate safeguards for the driver.

Every year operators are injured through careless handling of chemicals. Some sprays are made more dangerous by their cumulative effect. It is the gradual build-up of poisonous chemicals rather than the immediate effect which does the damage.

The Health and Safety Executive has leaflets which give guidance on regulations and advice on precautions. With some chemicals certain protective clothing must be worn.

The obvious precautions, such as not spraying on windy days, particularly if the wind is blowing towards livestock, or near pastures in use, susceptible crops, gardens or orchards, need an annual reminder. Read the chemical tin label carefully; don't rely on memory. The makers may have changed its type and concentration and the application rate.

Send back the empties without delay if they are returnable. Burn or wash out, puncture and bury the others. Lock up those stored and don't keep them near fertilisers, feed or seed.

Wash the sprayer with detergent — inside and outside — immediately after use. Don't let the washing water seep or drain into domestic or other water supplies.

Operators should wear protective clothing, never try to clear a blocked nozzle by sucking or blowing, and never smoke while working. They should wash well before smoking, eating or drinking, and after work. Keep clear of spray drift, wash and maintain protective clothing before storing.

Pumps

Three main types of pump — gear, roller vane and diaphragm — are in common use on the farm. All will apply popular chemicals, but if gritty suspensions or liquid fertilisers are to be used in any quantity the following information will be useful.

Gear pumps are fast disappearing from regular use because they wear quickly with suspensions. The pump has two gears, one driving, one driven, contained in a tight-fitting metal casing with inlet and outlet ports. An adjustable relief valve in the circuit prevents overloading and controls delivery pressure.

Liquid is carried round the outside between the gear teeth and the casing and delivered under pressure as the teeth come into mesh.

Gear pumps are suitable for non-suspension chemicals only as grit will rapidly wear teeth, end plates and casing. Gears and housing are made of phosphor bronze and/or steel and the main point to check is wear between gears and end plate. Place a straight edge across the face of the gears and casing and measure the gap between the straight edge and gear faces with a feeler gauge. Any more than 0.025 to 0.040 mm will allow liquid

Fig. 6.105 Watch for wear on the gear pump cover plate. Replace or have it faced if scored.

to escape and the required pressure will not be reached (Figs. 6.104 and 6.105).

Roller vane pumps are a fairly common type, being fitted to many sprayers. Their main advantages are that relatively high pressures can be reached and they have a long life with proper maintenance. A cast-iron chamber has an offset slotted rotor, and rollers held in the slots roll in and out under centrifugal force, carrying liquid from inlet to outlet point. Liquid is forced out under pressure as the gap between rotor and housing decreases opposite the delivery port.

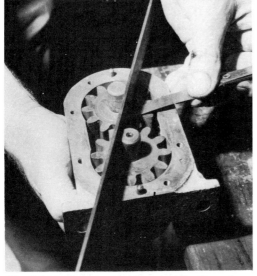

Fig. 6.104 Gear pumps wear rapidly if used with suspension chemicals. Check condition of the gear faces with feeler gauges as shown above.

Fig. 6.106 A roller vane pump should be checked once a year. Replace all the rollers if any are cracked or show signs of ridging.

Two types of roller are common, a weighted nylon one for non-gritty chemicals and a synthetic rubber or ebonite roller for gritty materials. Carry a spare set of both types of roller and spare sealing gaskets so that it is easy to change from one type of chemical to another.

Inspect the pump once a year for damaged rollers. Worn rollers will be cracked or ridged and will give low pressure. Replace them in sets to maintain the pump's balance (Fig. 6.106).

Diaphragm pumps, although initially more expensive than gear or roller vane pumps, have a long life and need little maintenance. All types of liquid can be applied with them. Most parts in contact with liquid are rubber or plastic, so there is no rust problem.

The pump works like a fuel lift pump in an engine. Main troubles are sticking valves or a holed diaphragm.

Maintenance is a simple matter of lifting the head and fitting a new rubber diaphragm and plastic valves. Most diaphragm pumps are multichambered to damp out pulsations inherent in the design (Fig. 6.107).

(c) Removing a diaphragm.

Always turn roller and gear pumps by hand before fitting to the pto, and prime them before use. Never run them dry or rapid wear will occur. Grease them regularly.

Store gear and roller pumps full of oil. If they are to be left unused for a few days run de-watering fluid through them to prevent rusting. Always tie a pump to some part of the tractor frame to prevent it rotating round the shaft and damaging the hoses (Fig. 6.108).

Fig. 6.107 Overhauling a diaphragm pump.
(a) Removing a valve.

Fig. 6.108 Always fix the pump body to the tractor to make sure that it cannot rotate.

(b) Exposing a diaphragm.

Keep all connections tight and when reassembling make sure all bolts on the face plates are tightened evenly to prevent distortion.

Markers for accurate spreading

With fertiliser, spinner spreading widths and spray boom lengths increasing, some form of marker to indicate where the last bout finished is essential.

Mr John Pope made a cheap set of markers from 40 mm diameter tube and 12 mm diameter round bar for the 12 m spread of his fertiliser distributor. He attached the markers to the tractor front weight frame (Fig. 6.109).

Two 6 m booms are made with the tube and braced, to prevent them bending and bouncing, with struts made from the round bar. The end of the bar that fits into a hole on the angle-iron carrying frame is threaded to take a nut.

Each boom is slid on a prong on the angle-iron framework and the threaded ends of the bars are pushed through two 16 mm holes in the upright part of the frame. A nut and old valve spring hold each boom in position and cushion shock loads.

Fig. 6.109 Each half of the boom is secured with one nut and a spring.

A chain of 1 m bolted to each end of the boom acts as a marker. To make more easily seen marks, Mr Pope ties a small coil of barbed wire to the end of the chain.

Fitting the booms is a one-man job. Rest one end of the boom on a short length of wood stuck in the ground while the nut and spring are fastened.

Dye marker for spreader accuracy

Mr Alan Kyle has come up with a simple inexpensive, home-made device that marks the ground with dye as the tractor drives across a field spreading fertiliser.

This has solved a problem for Mr Kyle who has been taking three silage cuts for several years, and for any farmer spreading fertiliser immediately after cutting silage. The white stubble makes it difficult for the tractor driver to see the tracks on the previous run.

Dye dripping on to one wheel of the tractor marks the stubble and enables the driver to get an even spread of fertiliser on his next run.

Using an 8 m oscillating spout, Mr Kyle finds 90 litres of diluted dye will mark about 3 hectares.

Crystal violet dye can be bought at the local chemist's shop. A heaped teaspoonful in 20 litres of water makes a suitable concentration.

The bottom part of an old can was mounted on the tractor and a 20 litre oil drum fitted with a tap was placed in it. A rubber hose from the tap drips the dye on to the front wheel. An open-shut valve is operated by a wire from the driver's seat. Figure 6.99 shows how to set the markers of a typical grain drill.

HEDGE CUTTERS

Regular inspection, knowing what to look for, the day to day sharpening procedure, and deciding when to send a blade to a saw doctor or timber man for attention, are equally important.

A copy of the appendix to the handbook giving the profiles and tooth specification of the blades should go with the blade for repairs. This will tell the saw doctor all he wants to know.

Blades should be inspected twice a day for cracks and tension, and always after hitting stones or metal or overheating. Test by sight and sound. Strike the blade with a piece of wood. Get to know the true 'ring' of your blade. A blade which has lost its tension or temper will ring at a lower note. A cracked blade will sound 'flat' and the ringing tone will not persist. If the note of the blade has changed do not use it until it has been passed as sound by a saw doctor.

Sharpen a little and often or it will become

a major operation for an expert.

Do not change the shape of the teeth or make sharp corners at the roots. Use as a guide the profiles in the appendix of the handbook. Note the curved gullets (Fig. 6.110).

Fig. 6.110 Badly maintained blades.

(a) An example of a badly maintained sawblade — the profile is incorrect, it has a broken tooth, and the square shows 'blueing' caused by overheating.

(b) This slasher blade has incorrect profile and worn cutting edges built up by welding. The heat of the welding flame will have caused the blade to lose its temper, the wrong profile has caused shock loading, and several teeth have broken off.

Use a round rat-tail file for the gullets. Avoid flat files or a flat disc on a power tool, as these will leave a sharp corner and create a stress point. Heating caused by the power tool will also draw the temper.

The hammering a blunt blade receives from the hedge, and loss of temper due to overheating, increases stress. If this is concentrated at a stress point a hair-line crack may develop.

With slasher blades make sure the tip of the tooth's cutting edge has not been worn away and become rounded. If necessary grind away the circumference of the blade, so as to make sure that the tip of the cutting edge — not the circumference of the blade — makes first contact with the wood. Use a slow-speed grindstone not a power tool — this avoids overheating.

It is important to remember that slasher blades have no 'set' and must never be given one.

With saw blades make sure that all the teeth are the same size and shape and have the same 'set' in accordance with the manufacturer's instructions. This is to ensure that no single tooth takes excessive load.

The correct set for the teeth of the McConnel Shapesaw saw blade, for example, is 0.6 to 0.75 mm. Depth of the set should be no more than 10 mm. Never start the set from the root of the tooth (Fig. 6.112).

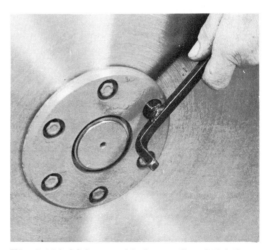

Fig. 6.111 Make sure blades are fitted tightly. Fit the blade, add the retaining plate and locking ring, fit bolts and tighten the nuts hard. Run the blade in a hedge for five minutes, stop and retighten the nuts. In some cases up to 1½ turns can be taken up after 'bedding in'.

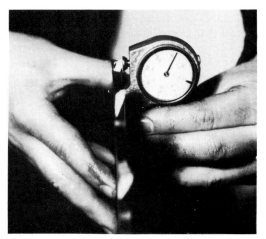

Fig. 6.112 This is the special gauge for checking the 'set' of a McConnel Power Arm Shapesaw blade showing the correct 'set' — 0.75 mm. This 'set' must not start at the root of the tooth and should never be more than 10 mm deep.

Lack of set leads to overheating and 'blueing' of the blade, which destroys temper. If this happens the blade must go back to a saw doctor for heat treatment 'hammering.'

The importance of the correct remedial treatment stems from the make-up of the blade. It is forged from a single billet of high alloy steel and not cut out of a sheet, as is popularly believed. This gives a radial flow to the grain in the steel. The blade is then cut out, heated to release stresses, hardened and tempered to the required toughness. After sharpening around the edge, it is 'hammered' by the saw smith to expand the centre but not the outer section, to put into the blade preloaded tension which stops it flapping at high speed. Finally, the surface is ground at low speed without heat to remove marks and enable it to be inspected.

The law allows one broken tooth to be filed down and the saw re-used, but it first needs inspection by a saw doctor to see that other hidden damage has not been caused by the blow which broke the tooth.

Many firms do not advise working with even one tooth short. Better, they say, to have all teeth removed, a new set recut all round and the blade re-treated. This is usually only half the cost of a new blade.

CHAIN SAWS

The two main signs that a chainsaw needs sharpening are a large deposit of gum on the top plates of the cutters and the production of sawdust rather than wood chips. Figure 6.113 illustrates the components of a typical chain and Fig. 6.114 shows the parts of an individual cutter.

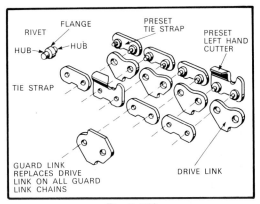

Fig. 6.113

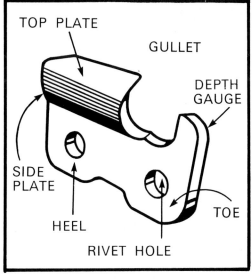

Fig. 6.114

Before sharpening, examine chain and cutters for damage. Then remove the chain and check it for tight joints. These are usually caused by a worn sprocket or the chain being operated when it is too loose. The stiffness is usually due to burred edges in the bottom of the tie straps. A chain that is too burred is irreparable.

Make sure that the chain has no cutters damaged beyond repair due to striking a stone or nail during work. A damaged cutter can be replaced fairly easily after carefully punching out the retaining rivets, preferably after grinding off the heads.

Before fitting new parts; file the bottom edges of the cutters and tie straps so that they match the rest of the chain (Fig. 6.115). When riveting on the cutter, mount it facing in the correct direction and be certain to place the tie strap on the chain so that the dimple or identification mark faces outwards (Fig. 6.116). This is because the tie strap is countersunk on this side (Fig. 6.117).

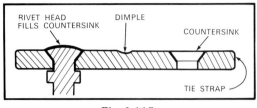

Fig. 6.117

Unless you are competent with small tools, have links and cutters replaced by a professional. He can replace several cutters in minutes and, with specialist equipment, make the joints much stronger.

A chain can be rejoined with new links quite safely but it is questionable whether a chain actually broken during work should be rejoined. It is liable to break again — and a chain breaking when rotating at 80 kmph is dangerous.

To sharpen the chain, hold it in a purpose-made filing vice, or between two lengths of wood in the workshop vice. It can be sharpened on the bar by holding the bar in a vice. But if this method is used, tension the chain correctly beforehand.

A file holder (Fig. 6.118) is essential. Few people can sharpen a chainsaw properly without one. Use the correct size of file, which will depend on the pitch of chain and the type of cutter.

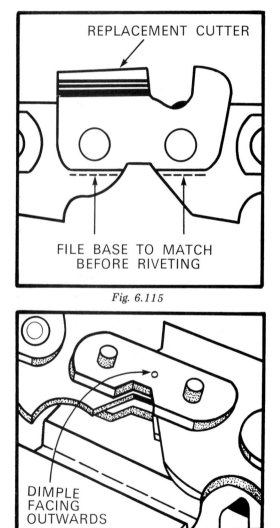

Fig. 6.115

Fig. 6.118

The pitch of the chain is determined by measuring the distance between any three rivets and halving the results.

The type of chain found on the farm will usually have a form of chipper or chisel cutters, but this must be checked.

A well-maintained chain can be useful, even when there is little left of the actual cutting blades.

Fig. 6.116

These counter-sinks are not obvious. After fitting a new link, check that the joint is not tight and file back the top plate of the new cutter to the same size as the others.

The worst cutter must be found and sharpened first. Afterwards, the length of its top plate must be measured and all the other cutters reduced to this length, otherwise the chain will cut sideways and wear the guide bar prematurely. A good tool for comparing the length of the cutters can be made from a fencing staple (Fig. 6.119).

Fig. 6.121 Correct way to hold the file.

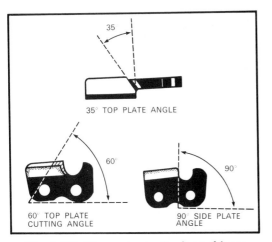

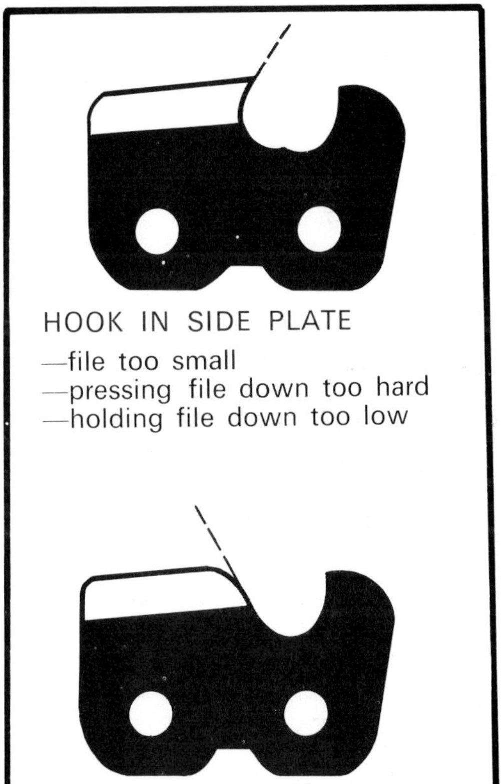

Checking length of cutter with a staple

Fig. 6.119

Cutters

If all the cutters are the same length before sharpening, count the strokes required to sharpen the worst cutter and give the other cutters the same number of strokes.

Find out which side of the chain is most difficult to sharpen and do this first before

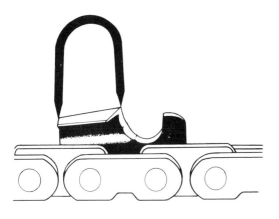

HOOK IN SIDE PLATE
—file too small
—pressing file down too hard
—holding file down too low

BACKSLOPE IN SIDE PLATE
—file too large
—holding file too high

Fig. 6.122

Fig. 6.120 Sharpening angles for a chipper chain.

35° TOP PLATE ANGLE

60° TOP PLATE CUTTING ANGLE

90° SIDE PLATE ANGLE

you start to tire. The file must be regularly rotated in its holder, because its teeth will soon lose their edge. Use a new file every time the chain is sharpened.

The sharpening angles for a typical chipper chain are shown in Fig. 6.120. These angles will vary with chain type and the wood cut.

To achieve the angles, first line up the file guide so that the 35° mark is parallel to the chain.

This mark must be kept parallel to the chain throughout the cutting stroke. Hold the file level, or better still, with the handle slightly lower by about 5° to 10° than the tip (Fig. 6.121). About one fifth of the file's diameter should protrude above the cutter. Figure 6.122 shows two common sharpening faults. Always file from the inside to the outside of cutter.

Fig. 6.123 Use of a gaugit.

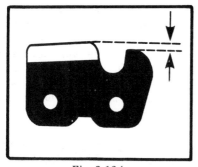

Fig. 6.124

After sharpening, tap the top of each cutter lightly with a wooden file handle to prevent the chrome plating on the teeth from peeling where it may have been damaged by sharpening.

Next check the depth gauges and lower where necessary. The normal height is about 0.6 mm (0.025 in) below the leading edge of the cutter. Do not file off too much from the depth gauges or this will cause the cutters to tip forward during work. Normally depth gauges only require filing every third or fourth sharpening.

Special tools to simplify depth gauge filing are available. Known as 'Gaugits' they are laid along the top of the chain and any part of the depth gauge which protrudes through the tool is filed off (Fig. 6.123). When all the depth gauges are level, round off the leading edges to restore the original shape (Fig. 6.124 shows the depth gauge setting).

The chain should now be removed from the bar and the drive link tangs cleaned with a round file where necessary to maintain a hook, which will clean the bar groove of sawdust. Then clean the chain and bar thoroughly to remove all traces of swarf and soak in oil before refitting.

The procedure will need to be modified slightly when dealing with chains which have different cutter angles. If hard or frozen wood is to be sawn, the depth gauges should be shallower and the top plate angle reduced.

As well as sharpening the chain, in order to achieve the maximum life from the chain, bar and drive system, the following maintenance is required.

Check the condition of the rails by placing a straight edge against one side of the bar to see that there is clearance between the bar and the chain (Fig. 6.125). When there is no clearance, the rails are badly worn and the bar will need replacing. Should the bar rails not be worn, mark the bar and remove it and the chain. The bar is marked so that it can be replaced upside down to wear both sets of rails evenly.

Once removed, the tops of the rails can be inspected for burrs and damage. Small burrs can usually be removed carefully with a file. Check that the slot between the rails is still deep enough to prevent the drive-links

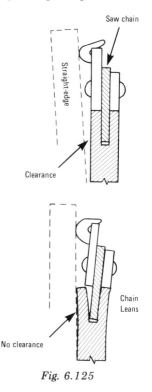

Fig. 6.125

Fig. 6.126 Cleaning the slot between the rails.

of the chain from touching the bottom. Clean the slot with a small screw driver or split pin (Fig. 6.126). Make sure the bar has not worn unduly in one spot — usually at the inner end of the bar — and check that the lubrication holes are clear. The drive sprocket must be in good condition or it will cause the chain to wear rapidly.

To change the sprocket, the centrifugal clutch assembly must be removed which means locking the crankshaft while the clutch retaining nut is loosened. Special tools are available. But the engine can sometimes be locked by tying a large knot in one end of a length of starter rope and inserting this through the sparking plug hole so that the knot becomes jammed between the piston and the cylinder head as the piston approaches top dead centre.

The correct tool for the job is usually a special pin which is screwed into the sparking plug hole and contacts the piston in the same way as the knotted starter rope.

The nut which retains the clutch normally has a left-hand thread to prevent it working loose when the engine is started. Always remove the sparking plug lead before carrying out any maintenance which may cause the engine to turn.

While the clutch is off, inspect and lubricate the bearings. Check the clutch linings although these should not show much sign of wear unless the saw has done a lot of work. Premature clutch wear is a sign that the saw has been operated at too low a throttle opening.

Next, check the bar and chain lubricating system. Most saws now have an automatic

oil pump and a manual system which is used to give the chain its initial lubrication. The oil pump usually ceases to work because of clogging.

The pump is housed behind the clutch which must be removed first.

The common types of oil-pump drive have either a worm gear or an eccentric. Both types are adjustable but professional advice is needed before adjusting the oil supply, especially when reducing it.

Always use a special chain oil, not engine oil. Chain oil is sticky and will not be thrown off the chain as it revolves. Contrary to popular belief it is not a good sign to see oil flying off the end of the chain when the saw is revved up.

Most oil tanks have a small filter which needs to be cleaned with petrol. Do not attempt to determine whether an automatic oil pump is working by revving the saw with the chain and bar removed. This can cause the engine flywheel keys to shear and the centrifugal clutch housings to become unscrewed. A clutch coming off at high speed can have fatal results. Sprocket or roller-nose bars require greasing through the hole shown in Fig. 6.127 every time the fuel tank is filled.

Fig. 6.127 Lubrication hole for roller nose bar.

After replacing the sprocket bar and chain, tighten the chain until its tie-straps just touch the bottom of the bar. The nose of the bar should be held upwards whilst this adjustment is made.

After tensioning, it should be possible to move the chain in its correct direction quite easily by hand (Fig. 6.128). Should the chain need readjusting afterwards and when hot, allow it to sag very slightly so that the tie straps do not quite touch the bottom of the bar.

Fig. 6.128 Moving the chain by hand after tensioning.

Never adjust the chain with the engine running.

SAVING FUEL

A few hours workshop time spent on a tractor on a rainy day can repay itself a hundred-fold in better fuel economy and great efficiency in completing the mammoth autumn cultivation blitz.

Running a tractor efficiently also involves how it is driven and worked. Where possible ensure that the tractor and implement are matched and that the hydraulic system of the tractor is used correctly to obtain maximum wheel grip. For example, select draft control, or one of the several variants, when using soil-engaging implements.

The diesel engine is an economical power unit but it does require maintenance to keep performance up to scratch. Often the decline in power due to neglect is so gradual that it remains undetected for some time during which plenty of fuel can be wasted.

To run efficiently, an engine must be mechanically sound. Its valves and piston rings must be in good condition in order to prevent loss of pressure from the cylinders during combustion.

Most of the routine maintenance required by a diesel engine does not involve a great deal of expenditure on spare parts and usually the saving in fuel will more than compensate for the time spent. Figures 6.129 to 6.139 will help guide the fuel-conscious driver through a fuel-thrifty routine.

Fig. 6.129 Every engine must 'breathe' properly. Service the air-cleaner more frequently when there is dust about. A partially choked air-cleaner can cause a 20% reduction in power.

Fig. 6.130 Incorrect valve-stem clearances (tappet clearances) will affect engine performance. Excessive clearance will prevent the engine from breathing and exhausting properly. Insufficient valve-stem clearance can eventually lead to burnt-out valves.

Fig. 6.131 Injectors which are producing poor atomisation cause extra fuel to be burnt due to inadequate mixing of the fuel and air. It pays to have the injectors checked or replaced at the recommended periods.

Fig. 6.132 Occasionally, fuel pumps are fitted with a manual retard start-screw. This must be screwed out once the engine is running or the fuel pump timing will be incorrect, resulting in loss of power and excessive fuel consumption.

Fig. 6.133 An engine must run at its correct temperature. An over-cooled engine will use excessive fuel. If the temperature gauge indicates that the engine is running cold, check the thermostat.

Fig. 6.134 Keep the tyres inflated to suit the job in hand. Excessive pressure will cause wheel-slip. Pressure which is too low will cause drag, but it may pay to reduce it slightly below recommendation.

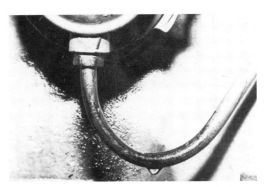

Fig. 6.135 There is no excuse for visible fuel leaks around unions. Each small leak probably wastes 10 litres to 20 litres a year. This can add up to a substantial loss of fuel.

Fig. 6.136 Exhaust pipes which are damaged or fitted upside down can reduce engine efficiency.

Fig. 6.137 The tracking must be correct or the tractor will use extra fuel in order to overcome the increased rolling resistance. Incorrect tracking also means that tyre wear will be unnecessarily high.

Fig. 6.138 Always remove excess weight. Keeping the front weights in position when they are not needed uses extra fuel.

Fig. 6.139 Check that the setting of the hydraulic system is fully understood, so that maximum benefit can be obtained.

CHAPTER 7

Plumbing

A leaking ball valve can waste considerable quantities of water.

The most common ball valve on the farm is the Portsmouth type (Fig. 7.1) which has a plunger moved horizontally by the action of a lever arm and float.

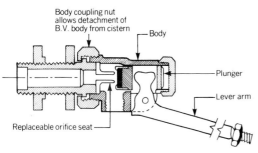

Fig. 7.1 Diagram of working parts of a Portsmouth type ball valve.

The plunger has a rubber seating which seals the inlet when the float is lifted by the water.

Ball valves for low pressure work (about 40 psi) will have bigger outlets and may carry the letters LP.

On some modern ball valves you can remove the inlet orifice and replace it with one of a different size so that the same body can be used for high and low pressure. The orifice seat can also be easily replaced when worn.

REPAIRING A BALL VALVE

Leaking float: A float with water in it will not be sufficiently buoyant to close the valve. A damaged copper float can usually be repaired by soldering.

Sticking plunger: This may occur in districts with hard water due to a build-up of mineral which eventually restricts the movement of the plunger. The remedy is to dismantle the valve and carefully clean off the deposits, taking care not to damage the surfaces of the plunger and valve body.

Alternatively, the salt build-up can be removed with a commercial de-scaling liquid. Diluted hydrochloric acid can also be used, but care must be taken to avoid accidents and damage to paintwork and clothing. Always use hydrochloric acid or a de-scaling liquid in a plastic or porcelain container. A metal container could be attacked by the acid.

Damaged or worn valve assembly: The valve seat eventually becomes worn due to age. When this occurs prematurely, the cause is probably the water affecting the brass seat and removing the zinc from the brass leaving a porous copper. This process is called de-zincification.

The best cure is to fit a ball valve with a nylon seat or one made entirely from plastic. A ball valve with a nylon seat will also have more resistance to wear caused by the rapid flow of water.

Provided the orifice seat is not damaged, a leaking valve can be cured by replacing the rubber seal in the end of the plunger. Turn off the water supply and remove the split-pin which retains the lever arm and also the end cap when fitted (Fig. 7.2). This allows the lever arm to be removed and the plunger to be pulled out. Take care not to drop them in the cistern.

Fig. 7.2 Remove split-pin which retains the lever arm.

Fig. 7.3 Unscrew the end of the plunger in order to remove the seal.

To replace the seal, the end of the plunger should be unscrewed (Fig. 7.3). But unscrewing it is usually so difficult that the plunger is damaged in the process. The alternative is to dig out the old one with a small sharp screwdriver (Fig. 7.4) and then to carefully push a new seal into the groove (Fig. 7.5).

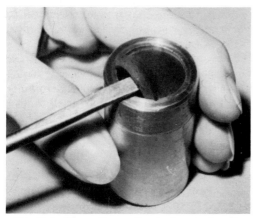

Fig. 7.4 Alternative method of removing the seal.

Fig. 7.5 Curl the new seal up.

In an emergency a new seal can be made from an old rubber Wellington boot. It will be neater when a hollow punch is used to cut it rather than a pair of scissors.

When the orifice seat has to be replaced, access is usually obtained by unscrewing the valve body (Fig. 7.6) and removing the seat from the rear of the body (Fig. 7.7). Be sure that the replacement seat is of the correct size for the water pressure.

Fig. 7.6 Unscrew the body in order to gain access to the orifice seat.

Fig. 7.7 Remove the orifice seat from the rear of the body.

When re-assembling the valve, use a brass, not a steel, split-pin. Steel will corrode and may cause a flood later.

Sometimes a new ball valve will leak because the float and its arm may require adjusting. The usual method is to carefully bend the lever arm downwards, taking care not to damage it. These arms are rather prone to snap off near the pivot point.

Special ball valves are available. The Garston valve, developed at the Building Research Station has a diaphragm to seal the inlet orifice. As all the working mechanism is housed on the dry side of the diaphragm, it is protected from attack by corrosive water and will not become stuck due to the salt deposits.

In another special ball, the valve plunger can be removed without turning off the water supply. It has a small shut-off poppet valve behind the valve seat. This extra valve is normally held open by the plunger, but as the plunger is removed it immediately closes behind the orifice and prevents water flowing while the plunger is removed refit a new washer.

The most common cause of a leaking tap is a worn washer. But when the valve seat in the tap body is damaged a new valve will not stop the drips and you will need to take further action.

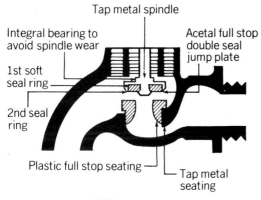

Tap metal spindle

Integral bearing to avoid spindle wear

Acetal full stop double seal jump plate

1st soft seal ring

2nd seal ring

Plastic full stop seating

Tap metal seating

Fig. 7.8 Pillar tap components.

Figure 7.8 shows the main components of a typical pillar tap. The mechanisms of most stop-cocks, bib and globe taps are very much alike and the repair procedure is similar.

FITTING A NEW WASHER

Check that the new washer is the correct size. Taps are the same size as the pipe on which they fit. With hot water systems, check that the new washer is made of a material which can withstand high temperatures.

Having turned off the water supply, unscrew the tap's shield. The capstan, or handle, may also have to be removed. Most taps for outdoor or commercial use do not have a shield and the head-gear will be visible.

The head-gear can be unscrewed from the body using a suitable spanner. When the head-gear is tight, the body of the tap must be supported while it is unscrewed to prevent the body rotating which can damage the pipe or fitting.

With the head-gear removed, the washer will be visible. Sometimes the jumper which holds

the rubber washer is attached to the head-gear, but usually it can be removed as in Fig. 7.9.

Fig. 7.9 Removal of jumper and washer.

The replacement washer is often fitted on to a new jumper. But the older type of flat washer has to be removed from the jumper.

The washer will be either a push-fit on the end of the jumper, or else retained by a small hexagonal nut. Some replacement washers do not resemble the original. A replacement washer which is mushroom-shaped and fitted with its own jumper can be used in place of most conventional types.

While the tap is dismantled, inspect the seat in the body. A seat worn by water erosion is unlikely to hold water even with a new washer.

There are two methods to repair a worn seat:

Re-cut the seat

The seats of taps and stop cocks can easily be re-cut using a seat re-cutter (Fig. 7.10). Select the correct size cutter and screw it on the end of the tool, wind the adjuster fully out and screw the tool into the body of the tap (Fig. 7.11).

Fig. 7.10 Tap seat recutting tool.

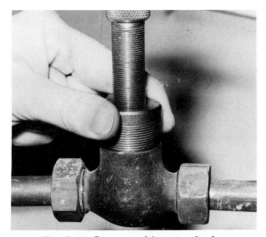

Fig. 7.11 Screw tool into tap body.

Fig. 7.13 Plastic seat kit.

Fig. 7.12 Rotate adjuster to lower cutter.

Fig. 7.14 Replace with new assembly.

Rotate the adjuster clockwise to lower the cutter on to the seat of the tap (Fig. 7.12). Rotate the handle clockwise a few turns in order to clean up the seat. Only a few turns are required to recondition a seat but, if necessary, the cutter can be adjusted to take a deeper bite and the handle given an extra couple of turns to complete the new seat.

Fit a plastic seat

Kits are available which consist of a new plastic seat to fit inside the old one and a plastic jumper complete with its own washer (Fig. 7.13).

Replace the old washer and jumper with the new plastic type (Fig. 7.14) then insert the tapered end of the plastic seat into the original seat.

The new seat is pressed fully home by re-assembling the tap and screwing the capstan, or handle, clockwise so that the jumper and washer are forced down on top of the seat.

These kits are simple to use and are much cheaper than a seat recutting tool. Normally they extend the life of a tap with a worn seat. But there are difficulties. The flow of water is reduced due to the smaller orifice in the plastic seat compared with the original orifice size. When the old seat is very worn, water will escape around the outside of the plastic seat which may be solved by applying a sealing compound to the outside of the new seat before it is fitted.

Before reassembly, check that the gasket between the headgear and the body is intact and sparingly lubricate the threads with petroleum jelly to ensure that it will be easy to dismantle.

Sometimes a tap will start to leak where the spindle protrudes from the head-gear because of a faulty gland seal. This can often be remedied by slightly tightening the gland nut. Do not overtighten this nut as the tap will become difficult to turn on and off.

Should tightening the gland nut fail to prevent the leak, then the gland requires re-packing. Unscrew the gland nut while holding the head-gear steady with another spanner.

Lift out the gland nut and remove the old

gland packing — usually a proprietary packing material or plumbers' hemp.

There may be a fibre washer underneath the hemp. Some modern taps may now be fitted with rubber o-rings instead of hemp. When a replacement o-ring of the correct size is not available, hemp can often be used instead.

To replace the hemp, a length of about 150 mm should be wrapped around the spindle in a clockwise direction (Fig. 7.15).

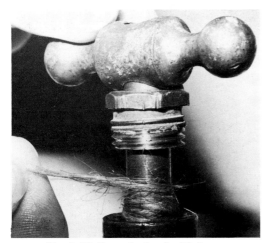

Fig. 7.15 Repack gland with hemp.

The gland nut is then tightened until the spindle is slightly stiff when turned.

Moistening the hemp with oil or grease will make it easier to fit and will also allow the gland nut to be screwed tighter before the spindle becomes difficult to turn.

Gland nuts of stop-cocks which are not frequently turned on and off can be screwed down fairly tight.

JOINTS IN COPPER PIPES

Copper pipes can be joined in many ways, most commonly with compression fittings and soldered joints. They can also be brazed.

Compression fittings

A typical compression fitting (Fig. 7.16) has a watertight seal created when the soft ring, known as an olive or ferrule, is squeezed between the outside of the pipe and the fitting. First the pipes must be cut square and their outsides cleaned with steel wool or fine glass-paper. Next a nut is placed on each pipe, followed by

the olives which have two chamfered faces, one of which is often longer than the other. Always fit the olives so that the larger of the two faces points away from the fitting.

Fig. 7.16 Parts of a compression joint.

Fig. 7.17 Two types of olive.

Figure 7.17 shows two different types of olive. Push each pipe into the fitting as far as it will go and scratch the pipe near the fitting. The scratch will show whether either pipe moves out of the fitting when the nuts are tightened. Some plumbers advise smearing a little jointing compound on to the olives while the joint is being made to help guarantee a watertight joint. Compression joints are easy to make and extremely reliable, but they have the disadvantage of being rather bulky and expensive.

Soldered joints

There are two common methods. Both rely on capillary action to move the solder into the joint.

Pre-soldered fittings contain a small amount of solder, usually in a ring (Fig. 7.18). Again the pipes must be cut square, then cleaned thoroughly with steel wool or glasspaper outside and in. Insufficient cleaning usually results in a leak. Apply a suitable flux to the outside of the pipes and inside the fitting, then press the pipes inside the fitting and rotate slightly to distribute the flux. Again a scratch mark on the pipes will show whether either pipe has moved before the soldering starts. Both the fitting and pipes are then heated with a blow lamp or gas torch, until a ring of solder appears around the edge of the fitting. Some plumbers place a little extra solder on the edge of the fitting as an added precaution.

End-fed fittings are assembled like the pre-soldered type, after cleaning and fluxing inside and out. Use the centre line of the fitting as a guide to determine how far the pipe is inserted. Once the joint has been assembled, the fitting and pipe should be heated enough to melt the solder which must be applied to its edge. Once the solder starts to melt, keep moving it all the way around the edge of the fitting until no more solder is drawn in between the fitting and the pipe and starts to build up around the edge. The completed joint should have a neat continuous line of solder around the edge.

Fig. 7.18 Pre-soldered fitting.

Bronze welded joints

The bell joint, as it is sometimes called, is similar in principle to the taft joint used to join lead pipes. Bell out the end of the pipes (Fig. 7.19) facing the direction of water flow. To make the bell, the end of the pipe should be annealed by heating to a dull red, then quenched in water. Alternatively, the end of the pipe can be heated and the punch used to bell out the end while it is still hot. When the pipe is quenched, do not point its opposite end to your face, as steam may blow out during quenching.

The bell should be big enough to allow 1.5 mm to 3 mm between the rim and the inside pipe. Thoroughly cleanse the surfaces to be joined, place the pipes into position and clamp

them if necessary. Brazing is easier if the pipes are vertical (Fig. 7.20). Use a brazing flux and a slightly oxidising flame. The size of the nozzle will depend upon the size of the pipes to be joined. As a guide, a No. 2 or 3 nozzle should be sufficient for this pipe. The brazing rod should be made of silicon bronze. Before brazing, pre-heat the whole joint, especially the end of the inner pipe. Then heat the end of the brazing rod and dip it into the flux; keeping the flame on the joint, apply the rod.

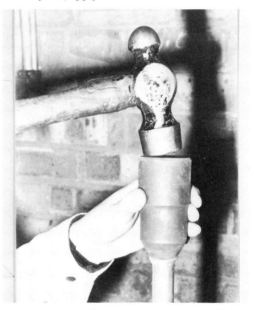

Fig. 7.19 Use a cone-shaped punch to bell the pipe end.

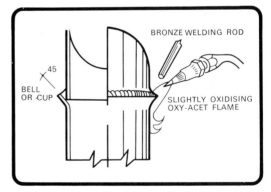

Fig. 7.20 Braze the joint vertically if possible.

When a small amount of rod has melted into the bell mouth of the joint, keep the heat on this spot so that the deposit spreads around the rim. Then add more bronze to the original spot in order to get an adequate build-up and progress around the rim.

When the job is going well, a small amount of molten bronze will precede the main deposit. The finished weld should be slightly built up and have a clean smooth appearance without any blow holes.

Do not overheat the pipes or excess bronze may run down inside and restrict the water flow. Overheating the pipes may cause the ends to melt. Do not quench the finished joint. Allow it to cool naturally. Copper is a good conductor of heat and the pipes could well be hot to touch some distance from the joint.

When a number of pipes have to be joined, it may be worth purchasing a special drift (Fig. 7.21). This can be used to expand one end of a copper pipe so that another pipe can be pushed

Fig. 7.21 Drift for expanding a pipe.

Fig. 7.22 Tool for grooving the inside of a pipe.

inside it to form an end-fed joint. Before this drift is used the pipe should be annealed. Figure 7.22 shows another special tool to fit inside a tube which has already been expanded. This tool usually fits two different sizes of pipe. When it is inserted in to a pipe and rotated, a ball bearing is forced outwards so that it forms a groove in the inside wall of the pipe. This groove can then be filled with solder so that the end of the pipe becomes a type of pre-soldered fitting.

PLASTIC PIPES

Plastic piping has transformed the laying of water supplies and waste disposal systems from time-consuming jobs for professional plumbers into relatively simple DIY tasks. Most plumbing difficulties can be solved cheaply and easily using a plastic component. For water supplies a further bonus is that plastic is not so easily damaged when the water freezes.

Permanent installation can be achieved by pulling the pipe into the ground behind a mole plough. Plastic pipes should not always be used as an underground water supply in close proximity to a gas pipe. Where it is necessary to do

Fig. 7.24 Make sure that the correct inserts and olives are used. These may need to be changed depending upon the type of pipe used. The picture shows two types of different inserts.

Fig. 7.23 The parts of a joint designed for use with polythene pipe.

so, first seek advice from the local water supply authority.

When the installation is intended to be permanent, always check that the pipe is suitable for the job in hand because there are many types of pipe and fitting.

Fig. 7.25 An insert prevents the tube collapsing inwards.

Fig. 7.26 Special connectors are available which join polythene pipe to conventional fittings.

Fig. 7.27 Underground waste disposal system made from plastic piping (fit a manhole at each sharp corner).

Fig. 7.28 Plastic pipes can be connected to salt-glazed pipes with special adaptors. This one is cemented to the outside of the salt-glazed pipe.

Fig. 7.29 A rubber sealing ring makes a watertight seal on waste disposal pipes. Don't forget to chamfer any sharp edges. Use washing-up liquid as a lubricant.

REPAIRING LEAKS IN STEEL PIPES

To repair a leak in a steel pipe either a new length of pipe must be fitted or else the offending section must be cut out and the pipe rejoined.

Rejoining steel pipe in situ is not straightforward because as one joint is tightened by rotating the pipe another one comes unscrewed. To allow a steel pipe installation to be repaired three different special joining arrangements are possible.

(a) *Pipe Coupling* This is made in two halves. Each half screws on to one of the pipes to be joined, then the coupling is joined using its own nut which clamps both parts of the coupling together (Fig. 7.30).

(b) *Running Joint* Basically such a joint involves threading one of the pipes with a long

parallel thread. A backing nut, sealing washer
and straight connector are then screwed on to
the parallel thread. The ends of the pipes are
correctly positioned and the connector screwed
back on to the second pipe and tightened in the
normal way. The backing nut is then tightened
down on to the connector. The washer between
the connector and the nut acts as a seal.
Plumber's hemp can be used instead of a sealing
washer (Fig. 7.31).

(c) *Johnson Coupling* This fits over the ends of
the pipes to be joined and is seated to the pipes
by tightening the nuts at either end (Fig. 7.32).
These joints are very quick and easy to fit.
They are intended to be temporary but most of
them will last for at least ten years.

Fig. 7.30 Pipe coupling.

Fig. 7.31 Running joint.

Fig. 7.32 A Johnson coupling being tightened
by rotating the end nut.

Metric Conversions

Conversions Metric ⟷ 'Imperial'/US

metric	'Imperial'/US
length	
1.0 mm	0.039 in
25.4 mm	1 in
305.0 mm	1 ft
914.0 mm	1 yd
1000.0 mm (1.0 m)	1.094 yd
1000.0 m (1 km)	1093.61 yd (0.621 mile)
1609.3 m (1.61 km)	1 mile
area	
1.0 cm^2	0.155 in^2
645.2 mm^2 (6.452 cm^2)	1 in^2
929.03 cm^2 (0.093 m^2)	1 ft^2
0.836 m^2	1 yd^2
1.0 m^2	1.196 yd^2 (10.764 ft^2)
0.405 ha	1 acre
1.0 ha	2.471 acre
1.0 km^2	0.386 mile^2
2.59 km^2 (259 ha)	1 mile^2
volume	
1 litre (1 dm^3)	61.025 in^3 (0.035 ft^3)
0.765 m^3	1 yd^3
1.0 m^3	1.308 yd^3 (35.314 ft^3)
capacity	
0.473 litre	1 pint US
0.568 litre	1 pint imp
1.0 litre	1.76 pint imp
1.0 litre	2.113 pint US
3.785 litres	1 gal US
4.546 litres	1 gal imp
mass	
0.454 kg	1 lb
1.0 kg	2.205 lb
0.907 t (907.2 kg)	1 ton US
1.0 t	0.984 ton imp
1.0 t	1.102 ton US
1.016 t (1016 kg)	1 ton imp

metric	'Imperial'/US
velocity	
0.025 m/s (25.4 mm/s)	1 in/s
1.0 m/s	39.4 in/s (196.9 ft/min)
1.0 km/hr	0.621 mile/hr
1.609 km/hr	1 mile/hr
temperature	
X°C	$(\tfrac{9}{5}X + 32)°F$
$\tfrac{5}{9} \times (X - 32)°C$	X°F

Technical Conversions

Torque
1 lbf ft = 1.356 N m
1 lbf in = 0.1130 N m
1 N m = 0.7376 lbf ft

Pressure and stress
1 lbf/in^2 = 6894.76 N/m^2
           =   0.0703 kgf/cm^2
1 kgf/mm^2 = 0.6349 tonf/in^2
1 kgf/cm^2 = 14.2233 lbf/in^2

Energy
1 ft lbf   =   1.8556 J (joule)
1 hp h   =   2.6845 MJ
1 Btu   =   1.055 kJ
1 therm = 105.506 MJ
1 kWh   =   3.6 MJ
1 calorie =   4.1868 J
1 J = 0.7376 ft lbf

Power
1 ft lbf/s   =   1.3558 W (watt)
1 hp   = 745.7 W
1 metric hp = 735.5 W

The joule (J) is the work done when the point of application of a force of one newton is displaced through a distance of one metre in the direction of the force (i.e. J = 1 N m); the watt (W) is the unit of power equivalent to one joule per second (W = J/s).

Appendix II –
Thread Tables

Table A1 Whitworth (BSW)

Size inch	T.P.I.	Tapping sizes		
		Inches	mm	Number or letter
$3/16$	24	$9/64$	3.6	27
$1/4$	20	$3/16$	4.7	12
$5/16$	18	$1/4$	6.4	E
$3/8$	16	$19/64$	7.5	N
$7/16$	14	$23/64$	9.1	U
$1/2$	12	$25/64$	9.9	X
$9/16$	12	$29/64$	11.5	—
$5/8$	11	$1/2$	12.7	—
$11/16$	11	$37/64$	14.7	—
$3/4$	10	$5/8$	15.9	—

Table A2 British Standard Fine (BSF)

Size inch	T.P.I.	Tapping sizes		
		Inches	mm	Number or letter
$7/32$	28	$11/64$	4.4	17
$1/4$	26	$13/64$	5.2	6
$9/32$	26	$15/64$	5.9	B
$5/16$	22	$1/4$	6.4	E
$3/8$	20	$5/16$	7.9	O
$7/16$	18	$23/64$	9.1	U
$1/2$	16	$27/64$	10.7	—
$9/16$	16	$31/64$	12.3	—
$5/8$	14	$17/32$	13.5	—
$11/16$	14	$19/32$	15.1	—
$3/4$	12	$41/64$	16.3	—

Table A3 Unified Course (UNC)

Size inch	T.P.I.	Tapping sizes Inches	mm	Number or letter
¼	20	$^{13}\!/_{64}$	5.2	7
$^5\!/_{16}$	18	$^{17}\!/_{64}$	6.6	G
⅜	16	$^5\!/_{16}$	7.9	O
$^7\!/_{16}$	14	—	9.3	U
½	13	$^{27}\!/_{64}$	10.7	—
$^9\!/_{16}$	12	$^{31}\!/_{64}$	12.3	—
⅝	11	$^{17}\!/_{32}$	13.5	—
¾	10	$^{41}\!/_{64}$	16.5	—
⅞	9	$^{49}\!/_{64}$	19.3	—
1	8	⅞	22.2	—

Table A5 British Standard Pipe (Gas) or BSP

Size of bore and thread inch	T.P.I.	Tapping sizes Inches	mm
⅛	28	$^{11}\!/_{32}$	8.7
¼	19	$^{15}\!/_{32}$	11.9
⅜	19	$^{19}\!/_{32}$	15
½	14	¾	19
⅝	14	$^{53}\!/_{64}$	21
¾	14	$^{31}\!/_{32}$	24.6
⅞	14	$1\,^7\!/_{64}$	28.2
1	11	$1\,^{15}\!/_{64}$	30.5

Table A4 Unified Fine (UNF)

Size inch	T.P.I.	Tapping sizes Inches	mm	Number or letter
¼	28	$^7\!/_{32}$	5.5	3
$^5\!/_{16}$	24	—	6.9	I
⅜	24	$^{21}\!/_{64}$	8.3	Q
$^7\!/_{16}$	20	$^{25}\!/_{64}$	9.9	—
½	20	$^{29}\!/_{64}$	11.5	—
$^9\!/_{16}$	18	$^{33}\!/_{64}$	13	—
⅝	18	$^{37}\!/_{64}$	14.5	—
¾	16	$^{11}\!/_{16}$	17.5	—
⅞	14	$^{13}\!/_{16}$	20.5	—
1*	12	$^{59}\!/_{64}$	23.3	—

* Note 1 inch ANF has 14 T.P.I.

Table A6 Metric

Size in mm	Pitch in mm	Tapping sizes Inches	mm	Number or letter
6	1	—	5	9
7	1	$^{15}\!/_{64}$	6	B
8	1.25	$^{17}\!/_{64}$	6.8	G
9	1.25	—	7.8	N
10	1.5	$^{21}\!/_{64}$	8.6	Q
11	1.5	⅜	9.6	V
12	1.75	$^{13}\!/_{32}$	10.2	Y
14	2	$^{15}\!/_{32}$	12	—
16	2	$^{35}\!/_{64}$	14	—
18	2.5	$^{39}\!/_{64}$	15.5	—
20	2.5	$^{11}\!/_{16}$	17.5	—
22	2.5	$^{49}\!/_{64}$	19.4	—
24	3	$^{53}\!/_{64}$	21	—
27	3	$^{15}\!/_{16}$	23.8	—
30	3.5	$1\,^3\!/_{64}$	26.5	—

Index